Anonymous

Contributions to Our Knowledge of the Meteorology of Cape

Horn

and the West Coast of South America

Anonymous

Contributions to Our Knowledge of the Meteorology of Cape Horn
and the West Coast of South America

ISBN/EAN: 9783337218539

Printed in Europe, USA, Canada, Australia, Japan

Cover: Foto ©berggeist007 / pixelio.de

More available books at **www.hansebooks.com**

CONTRIBUTIONS

TO OUR

KNOWLEDGE OF THE METEOROLOGY

OF

CAPE HORN

AND THE

WEST COAST OF SOUTH AMERICA.

Published by the Authority of the Meteorological Committee.

LONDON:

PRINTED BY GEORGE EDWARD EYRE AND WILLIAM SPOTTISWOODE,
PRINTERS TO THE QUEEN'S MOST EXCELLENT MAJESTY.
FOR HER MAJESTY'S STATIONERY OFFICE.
PUBLISHED BY E. STANFORD, CHARING CROSS.

1871.

[*Price* 2*s.* 6*d.*]

PREFACE.

THE collection of the information contained in the accompanying Charts was commenced under the direction of the late Admiral FitzRoy by Mr. G. H. Simmonds and Mr. F. Gaster. When the Meteorological Committee took charge of the Office it was not thought advisable to continue the collection of observations on the plan previously adopted, but at the same time it was resolved to turn to account the work which had already been done.

The observations are of good quality, but insufficient as regards quantity. It has, therefore, been an object of importance to give the information in such a form as to be capable of combination with other data when available.

The method adopted in construction of the Charts has been devised by Captain H. Toynbee, the Marine Superintendent of the Office, and the final discussion of the observations and preparation of the Charts for publication has been effected, under his directions, by Mr. R. Strachan.

It is hoped that the result will prove acceptable to navigators, whose requirements are the first consideration of the Office.

ROBERT H. SCOTT,
Director.

July 1871.

27587.

INTRODUCTION.

A key diagram and explanation are inserted on each chart, so that a few words will suffice to serve as an introduction.

It will be seen that the data given on the charts consist of barometrical readings, temperature of the air and sea, wind, and rain or hail, &c.

All instrumental readings have been made with verified instruments, duly corrected.

Barometer.—The barometrical readings have been reduced to 32° Fah., and to mean sea level.

In comparing actual observations made at sea with those given on these charts, it must be remembered that by reducing all readings to the uniform temperature of 32° F., the uncorrected readings taken at high temperatures have been much reduced. Thus, if the actual reading be 30·000 in., it will be found that

> at a temperature of 47° the amount to be subtracted is 0·050 in.
>> ,,   ,,   66° , ,,   ,,   ˙0·101 in.
>> ,,   ,,   85°   ,,   ,,   0·150 in.

Further particulars on this subject will be found in the Barometer Manual.[*]

The necessity for this correction is caused by the fact that the mercury in the barometer is expanded by heat, so that the same bulk of mercury is lighter the higher is its temperature. Accordingly it is only by reducing all the observations to the same temperature that a true idea can be given of the relative pressures at different points on the earth's surface. The temperature which is always selected is 32° F., the melting point of ice.

The hours at which the observations which have been used have been taken are usually

---

[*] Barometer Manual, Board of Trade, 1871. Potter, Poultry ; and Stanford, Charing Cross. Price 1s.

27887.   A

4 a.m., noon, and 8 p.m. Whenever observations at these hours were not available, the hours have been chosen which would give the best attainable mean for the day.

Wind.—The force of the wind has been given according to Beaufort's scale, and calculated to one place of decimals. One reason for adopting this plan is that a wide range exists between some of the consecutive numbers on that scale, *e.g.*, No. 4 is a wind force which will cause a "well-conditioned man-of-war" to go five or six knots an hour, while No. 5 is a force to which the same ship can just carry royals "on a wind," and may be going 10 knots or more.

It must be remembered that the figures in the inner circle only show the *mean* force of the wind, and do not give any information as to the *extreme* forces registered from each point during the month.

It is therefore necessary for the reader to consult the weather diagrams (pp. 20–25) to ascertain the number of squalls recorded during the month in that square.

The direction of the wind is given on the charts to 16 points only, and as in the original logs it is given to 32 points, it is necessary to explain how these observations have been grouped.

The directions were first corrected for magnetic variation. Then (the points being numbered from N. round by E., S., and W. to N. again, from 1 to 32, so that N. by E.=1,

E.=8, S.=16, W.=24, and N.=32) all the *odd* points were thrown back to the next *even* point, thus N. by E. was classed with N., N.E. by N. with N.N.E., and so on. By this means they were grouped as is shown in the diagram.

The result is, that looking from the centre of the circle towards any given point on the circumference, the absolute mean direction, strictly speaking, is half a point to the right of that shown on the key diagram on the map, when the direction has been as frequently recorded on the *odd* as on the *even* points; but as fair winds are more commonly recorded on *even* points than on *odd*, the charts may be considered to give a fair representation of the truth.

When any of the observations have been taken in port, the name of the place is inserted on the chart.

Weather.—It will be seen that the only information given on this head in the charts is that relating to rain, snow, or hail. Many logs contain entries of weather only when it is exceptional, the column being left blank when it is fine. Hence we find some difficulty in dealing with the subject; *e.g.*, in January, in the square from 40° to 45° S., and from 75° to 80° W., there are only seven observations of weather, five of which are rain, while there are 15 observations of the barometer and of wind. It is fair to suppose that if the weather had been in any way remarkable the fact would have been noticed in the log when the wind observations were entered, so that we may reasonably allow that five represents the number of times rain was observed, not in seven but in fifteen observations.

The above supposition has been made in each case in the remarks on the separate charts, and the number of rain observations has been referred to the total number of observations of whatever element has been recorded most frequently.

Isobaric Lines.—In drawing the isobaric lines, a line has been given for each tenth of an inch for 30 ins. and upwards. Below 30 inches they have usually been drawn at intervals of two tenths of an inch. But this rule has not been invariably followed. It is important to remember this, because the mean strength of the wind is related to the closeness of the isobars, and if these latter had been drawn for each tenth in the southern part of the charts, the appearance of the charts would have been very different from that which they now present.

Isothermal Lines. — The isothermal lines are drawn for every five degrees of air temperature.

Where there are no data blank diagrams have been given, so that the navigator may, if he wishes, fill them in with his own observations to serve for his future guidance.

The outline of the land has been given as faintly as possible to prevent its interfering with the data.

JANUARY.

Pressure.—The isobar of 30·0 passes northward at the meridian of 75° W. from 38° S. to 28° S., thence to 13° S. it appears to overlap the continent.

Those of 30·1, 30·2, and 30·3 are entirely upon the ocean, the area of greatest pressure being in 95° W. and 35° S. The district of low pressure with variable winds off Valparaiso is remarkable.

The isobar of 29·8 is in the mean latitude of 48° S.
    „   29·6   „   „   52° S.
    „   29·4   „   „   55° S.

Winds.—The south-east Trade wind extends at least to 30° S. With the exception of a few northerly winds off Valparaiso, a general southerly wind is felt along the coast from 40° S. to 20° S. Between the parallels of 40° and 60° the prevalent winds are westerly, but to the south of Cape Horn easterly and northerly winds are not unfrequently reported.

Temperature.—The isotherm of 70° takes a course like an S from 35° S. 95° W. to 17° S. 75° W.

„   65° trends gradually northward from 40° S. 92° W. to 26° S. 70° W. off the coast.

In lat. 35° the air is 10° warmer in 95° W. than it is off the coast.

„   60° near the coast is only slightly diverted to the northward from 42° to 37° S.

„   55° lies nearly on the parallel of 44° S.

„   50° bends southwards across Patagonia, nearly to 55° S., when it turns to the north again in the South Atlantic, indicating that the air on the mainland of Patagonia is warmer than that over the sea, as might be expected in the height of summer.

„   45° is in latitude 57° S.

Sea Temperature.—The sea temperature to the east of Cape Horn is about 2° colder than that of the air, but in the square lying to the south-east of that point, between 60° and 65° W., it is as much as 7° below the air temperature.

Rain—Is recorded between 60° and 55° S. once in every 2 observations.

„	„	55°	„	45°	„	3	„
„	„	45°	„	40°	„	5	„
„	„	40°	„	35°	„	14	„
„	„	35°	„	30°	not at all.		
„	„	30°	„	25°	once in every 11 observations.		
„	„	25°	„	20°	„	5	„
„	„	20°	„	10°	„	8	„

FEBRUARY.

Pressure.—The highest pressure lies in lat. 32° S., long. 87° W. The isobars of 30·2, 30·1, and 30·0 sweep round this region, the last-named curve passing nearly parallel to the coast, at some distance from it.

The southern isobars lie generally in an east and west direction, and pressure on the whole is lower in the south than in the preceding month.

Winds.—The Trade wind still extends to 30° S., and the southerly wind along the coast appears at the parallel of 40° S. From this to 60° S. the winds are generally westerly, but they do not appear to blow so steadily as in January.

Temperature.—The isotherm of 70° runs in a N.N.E. direction from 34° S. 95° W. to 10° S. 80° W.

,,   65° runs on the parallel of 37° S. from 95° W. to 82° W., and then turns to the N.N.E. and cuts 25° S. in 72° W.

,,   60° and 55° are both nearly on the respective parallels of 40° S. and 43° S., and do not turn to the northward, so that while in 95° W. the isotherms of 60° and 70° are 480 miles apart, the distance between them in 75° W. is 1,680 miles.

,,   50° is in the latitude of 47° S. in 85° W., and it sinks to that of 53° S. over Patagonia, and then holds an easterly course to the Falkland Islands.

    South of this isotherm the temperature is very uniform.

Sea Temperature.—The number of observations is scanty, but it may be noticed that along the coast from 75° to 85° W. the sea is from 1° to 2° warmer than the air. Between the parallels of 35° and 40° S., and from 85° to 95° W., the sea is 3° warmer than the air, while in the square to the south-east of the Falkland Islands it is nearly 4° colder than it.

Rain is recorded most frequently to the westward of Patagonia, where it is noticed twice in every three observations.

Between 60° and 40° S. it occurs once in every three observations.

,,	40°	,,	35°	,, ,, ,, six	,,
,,	35°	,,	25°	hardly at all.	
,,	25°	,,	20°	it occurs once in every eight	,,
,,	20°	,,	10°	,, ,, ,, nine	,,

MARCH.

Pressure.—The highest pressure is in 37° S. 97° W. The curves of 30·2 and 30·1 appear to sweep round this spot. The isobar of 30·0 runs eastward to Valparaiso, whence it turns to the north, and sweeps nearly parallel to the coast till it ultimately turns to the S.W. again in 12° S. 87° W.

The isobars of 29·5 and 29·2 appear to be thrown unduly northwards by exceptional weather.

Winds.—The Trade wind ends at 25° S., and from that to 40° S. the southerly wind is noticeable along the coast. Round Cape Horn the wind appears to be from N.W. in the West to W. and W.S.W. in the South, conforming more or less to the isobar of 29·2 in. Easterly winds are occasionally met with to the south of the cape.

Temperature.—The isotherm of 70° commences in 26° S. 85° W., two degrees to the northward of its position in February. The course of this curve and of those of 75° and 80° appear to show that the effect of the cold polar wind along the coast extends as far as to the equator. The latter isotherm, that of 80°, is only shown in this month, evidently owing to the absence of observations in January and February.

    ,,   65° is nearly in the same position as in February.

    ,,   60° turns slightly to the north on approaching the coast.

    ,,   45° turns quickly southwards, reaching the parallel of 54° S. in 82° W., and even that of 56° S. off Cape Horn.

In the southern part of the chart the temperature nowhere falls below 40°.

The air over Western Patagonia is colder than that over the sea further west, as is shown by the course of the isotherm of 45°.

Sea Temperature.—The sea in 50°–55° S. and 70°–75° W. is 4° warmer than the air, and a similar difference, though in a less degree, extends all up the coast as far as to 30° S. and to the meridian of 80° W.

Rain is most frequent to the west of Patagonia, where it occurs about once in every two observations, between 55° and 60° S.

    Between 55° and 45° it occurs once in every 3 observations.

,,	45°	,,	40°	,,	,,	,,	4	,,
,,	40°	,,	35°	,,	,,	,,	9	,,
,,	35°	,,	30°	,,	,,	,,	6	,,
,,	30°	,,	25°	,,	,,	,,	25	,,
,,	25°	,,	20°	,,	,,	,,	7	,,

    ,,   20° ,, equator it was hardly noticed at all.

APRIL.

Pressure.—The distribution of pressure in this month shows a marked contrast to that which prevailed in the three previous months. It is now more uniform between 10° **and**

40° S. than it was, but the isobars take very sinuous courses. Two distinct patches of pressure above 30·2 are observable, but it will be noticed that one of these depends on only three observations.

Around Cape Horn itself there is a district with readings below 29.0 in. Here also the number of observations is so small, that the lowness of the pressure may be due to exceptional weather.

On the whole there is a general diminution of pressure on approaching the coast, between 10° and 40° S., and between 50° and 60° S. the mean pressures vary considerably.

Wind.—We still trace the southerly wind along the coast from 10° to 40° S., but its force is lighter between 30° and 40° S., as might be expected from the greater uniformity of pressure.

The limit of the S.E. Trade is in 25° S.

Southward of the Trades the wind is remarkably unsteady. It appears to be northerly in 30° S., between 90° and 100° W.

Easterly winds are not unfrequent to the south of Cape Horn.

Temperature.—The isotherm of 70° has moved still further to the northward, and runs from 27° S. 100° W. to 10° S. 80° W.

        ,,   65° runs parallel to it, at a distance of about 750 miles.

        ,,   60° is similarly parallel, at 350 miles distance from that of 65°.

        ,,   45° lies between 50° and 55° S. to the westward of Cape Horn, but it cannot be traced to the eastward as it lies north of 50° S., the limiting latitude of the district now being discussed.

        ,,   40° is in 58° S. off Cape Horn, but on the western side it passes south of 60° S.

Sea Temperature.—This scarcely differs from that of the air, so that it does not call for special remarks.

Rain.—The frequency of rain decreases with the latitude.

Between 60° and 55° S. it occurs once in every 2 observations.

,,	55°	,,	45°	,,	,,	3	,,
,,	45°	,,	35°	,,	,,	4	,,
,,	35°	,,	30°	,,	,,	6	,,
,,	30°	,,	25°	,,	,,	1ī	,,
,,	25°	,,	20°	,,	,,	14	,,
,,	20°	,,	10°	it is very rare.		

MAY.

Pressure.—The mean pressure varies between 30·0 and 30·3 all along the coast, and there are three separate areas where the readings are above 30·2. On the coast of Peru itself a slight increase of pressure is shown, for the isobar of 30·1 has taken the place of that of 30·0. There are, on the whole, less extremes of pressure than in the previous months, though in the higher latitudes the course of the isobars is most irregular, and the barometrical readings appear to fluctuate very much. The mean pressure near Cape Horn is about half an inch higher than it was in April.

Wind.—The Trade wind only extends to 20° S., and the southerly winds appear on the coast down to 45° S., though not so decidedly as in the former months. West of the meridian of 80°, and between 20° and 50° S., the winds are remarkable, but the observations are few in number.

Between 50° and 60° S. the winds vary greatly, and easterly winds are not uncommon. In some squares off the coast of Chili light winds, variables, and calms are reported, especially between 30° and 35° S. In the square from 40° to 45° S. and from 75° to 80° W. a high mean force is shown for the wind, but calms are also commonly experienced.

Temperature.—The isotherms of 75° and 70° maintain their north-easterly course, but have approached closer to the equator.

The isotherm of 65° lies approximately in 20° S.

,, 60° stretches to the E.N.E. from 37° S. 92° W. to 26° S. 70° W.

The remaining isotherms, those of 55°, 50°, 45°, and 40°, run nearly in an east and west direction in latitudes 40°, 42°, 48°, and 58° respectively, so that the decrease of temperature between the parallels of 40° and 50° is much more rapid than in higher latitudes.

Sea Temperature.—The sea is generally warmer than the air in this month, to the extent of 3° between 25° and 30° S., and of 2° between 35° and 30° S., while in the district between the meridians of 80° and 85°, and the parallels of 35° and 50° S., it is as much as 4° warmer.

Rain.—Between 60° and 40° S. rain is reported once in every 2 observations.

,,	40°	,,	30°	,,	,,	5	,,
,,	30°	,,	25°	,,	,,	14	,,
,,	25°	,,	20°	,,	,,	5	,,
,,	20°	,,	15°	,,	,,	15	,,
,,	15°	,,	10°	,,	,,	19	,,
,,	10°	,,	5°	,,	,,	13	,,

9

JUNE

Pressure.—There is very little information for the lower latitudes. The highest readings on the chart fall below 30·2 in. and are situated between 20° and 30° S.

South of this district pressure decreases rapidly and the readings in the southern squares are uniformly very low. The meridional direction of the isobars between 75° and 90° W. is very remarkable, and it will be noticed that pressure increases as you approach the land, being the reverse condition to what obtains in January.

Wind.—The Trade winds extend to 30° S., but they are very light and variable in some squares.

Southerly winds are still predominant along the coast of Chili and Peru, but are more frequently interrupted by winds from the opposite quarter than in the former months.

Along the belt of latitude from 30° to 35° S., there is a general westerly wind, in marked contrast to the Trade winds close to it.

Easterly winds are much more frequent in high southern latitudes.

Temperature.—There are no data for the isotherm of 70°.

The isotherm of 65° runs to the N.E. from 27° S. 87° W. to the coast.

       ,,   60° runs parallel to it at a distance of about 600 miles.

       ,,   55° runs parallel to the last and only 250 miles from it.

The curves of 50°, 45°, and 40° lie at a distance of about five degrees of latitude from each other and take an east and west course.

The isotherm of 35°, the lowest we have yet noticed, makes its appearance in 57° S. off Cape Horn.

Sea Temperature.—This is still higher generally than that of the air, but the difference is not so great as in May.

Rain is most frequent between 55° and 60° S.

    Between 60° and 35° it occurs generally once in every 2 observations.

,,	35°	,,	30°	,,	,,	5	,,
,,	30°	,,	25°	,,	,,	8	,,
,,	25°	,,	20°	,,	,,	24	,,
,,	20°	,,	15°	,,	,,	7	,,

No rain has been recorded north of 15° S.

JULY.

Pressure.—The maximum pressure on the chart is 30·4 in 33° S. and 92° W. In this portion of the chart the curves take a very sinuous course. The readings average about

27887.   B

29·6 in lat. 48° S.   South of that parallel they vary so much that it is not possible to draw the isobars so as to represent the facts in an intelligible manner.

Wind.—The Trades extend to 25° S., with frequent interruptions near the land between 20° and 25° S., thence to 35° S. the winds seem conflicting, but the belt of westerly winds noticed in June, between 30° and 35° S., is perceptible in the same locality.

The southerly wind along the coast prevails still, but it is more frequently interfered with by northerly winds.

The proportion of easterly wind in high southern latitudes is less than it was in June.

Temperature.—The isotherm of 70° stretches from 22° S. 100° W. to 5° S. 82° W.
,,   65° runs parallel to it, at a distance of about 300 miles
,,   60° runs E.N.E. to the coast, from 33° S. 100° W.
,,   55°, 50° and 45°, take an east and west course, at a distance of about five degrees of latitude apart.
,,   40° lies far to the southward in 55°.

The northerly deflection of the isotherms on the coast is not so apparent, a fact probably due to the diminished frequency of the polar winds.

Sea Temperature.—The sea is warmer than the air to the extent of 2° or 3° between 30° and 40° S., and also south of Cape Horn.

Rain.—This is now less frequent in the high latitudes, and more common between the tropics than it was in June.

Between 60° and 30° S. rain is recorded in general, once in every 3 observations.
,,   30° ,, 25°   ,,   ,,   8   ,,
,,   25° ,, 15°   ,,   ,,   7   ,,
,,   15° ,, 5°   ,,   ,,   4   ,,
And at the Galapagos Islands ,,   ,,   8   ,,

AUGUST.

Pressure.—Barometrical readings range above 30 in. between 10° and 45° S.   There are two distinct areas of maximum pressure between 30° and 40° S., in one of which the mean recorded is as high as 30·558; but this figure has been deduced from three observations only, and was accompanied by a polar wind and a low temperature, so that it carries but little weight as an average.

From the parallel of 45° to that of 53° S. the mean difference of readings is half an inch, giving a steep gradient for westerly winds, which in this part of the chart reach a mean force of about 7.   To the south of Cape Horn the isobar of 29·2 is met with, and the gradient to it from the next isobar of 29·5 is also steep, while the average force of the wind is as much as 8.

Wind.—The Trade wind seems to extend to 25° S. On the coast of Chili, between 30° and 40° S., the winds are very light, their direction ranging between S. and N. by the W. From 40° to 60° the winds are generally from W.N.W., and are strong. In high latitudes, easterly winds seem to be rare during the month.

Temperature.—The isotherm of 65° lies nearly in the same position as that of 70° did in July, and so takes a N.N.E. course.

The isotherm of 60° lies nearly in the parallel of 30° S.

The isotherms of 55°, 50°, and 45°, lie at unequal distances from each other, between 30° and 45° S. Their mean direction is nearly east and west.

The isotherm of 40° runs in latitude 53° for 15° of longitude, then dips southwards to 58° S. in 82° W., and from that runs to the N.E. on the western side of the Straits of Magellan, so that at Cape Horn and to the south and east of it the temperature does not reach 40°.

Sea Temperature.—This scarcely differs from that of the air, except to the east of Cape Horn, where the water is about 2° warmer than the air.

Rain.—This is most frequent in the south,

From 60° to 50° the frequency is	1 in	2
,, 50° ,, 40°	,,	,, 4
,, 40° ,, 35°	,,	,, 5
,, 35° ,, 30°	,,	,, 12
,, 30° ,, 25°	,,	,, 7
,, 25° ,, 20°	,,	,, 18
,, 20° ,, 10°	,,	,, 4

and close to the equator it is also very often reported.

SEPTEMBER.

Pressure.—The isobars have a much more regular course than in August. The area of high pressure in 90°—95° W. has reappeared, the readings in that district, between 25° and 35° S., being 30·4, and considerably higher than those near the coast.

In the high latitudes readings are very low, with strong westerly winds.

Wind.—The Trade wind extends in places to 35° S., and the southerly wind has again made its appearance on the coast.

Hardly any easterly winds are reported off Cape Horn, but this may partly be due to the fact that there are fewer observations than there were in August.

The northing in the westerly winds is not so conspicuous in the temperate zone as it was, and there seems a general tendency in the wind to circulate round the coast of Patagonia.

Temperature shows a marked increase when compared with that of the previous month. The isotherm of 70° is again seen in 13° S. and between 90° and 100° W.

The isotherms of 65° and 60° are at some distance from each other, and are each deflected to the northward along the coast.

Those of 55° and 50° conform much more closely to the parallels of latitude.

A remarkable bend to the northward is seen in the isotherm of 45°, off the west coast of Patagonia. This is caused by the low temperature, associated with southerly winds, in 80° W. 50° S.

The isotherm of 40° lies in latitude 56° S. up to 60° W., when it turns northward and passes just south-east of the Falkland Islands, rising to 52° S.

Sea Temperature.—The sea is slightly warmer than the air along the coast of Chili from 30° to 45° S.

Rain.—The frequency of rain is greatest in high latitudes.

Between 60° and 55° it is recorded twice in every 3 observations.

,,	55°	,,	35°	,,	once	,,	3	,,
,,	35°	,,	10°	,,	,,	,,	5	,,

Hence it appears that the rain is very uniformly distributed this month.

OCTOBER.

Pressure.—The highest pressure recorded is about 30·3 in., and it is situated in 30° S. and about 90° W.

The isobar of 30·2 envelopes this area, but without actually touching the coast. It descends as far south as to 46° S. in 83° W.

Below the parallel of 40° the mean pressures in the different squares vary so very much that it is impossible to draw the isobars.

Wind.—The Trade wind extends to 30° S., and closely connected with it, as usual, is the southerly wind along the coast, reaching as far as to 40° S.

From 50° to 60° S. the prevalent winds are westerly, a few observations of easterly winds being recorded south of Cape Horn.

Temperature.—The isotherm of 65° is nearly in the same position as in September.

,, 60° runs on a N.E. course from 35° S. 100° W. to 22° S. 75° W.

The other isotherms of 55°, 50°, 45° and 40°, are much less deflected towards the equator, especially in the higher latitudes, that of 40° lying nearly on the parallel of 57° S.

Sea Temperature.—There is a marked deficiency of observations of sea temperature. It appears, however, that the sea is on the whole colder than the air, except in the district lying between 30° and 35° S., and 90° to 95° W., where, however, the result seems somewhat suspicious.

Rain.—This is most frequent in the south :

Between 60° and 50° S. it occurs twice in every 5 observations.

,,	50°	,,	40°	,,	once	,,	4	,,
,,	40°	,,	35°	,,	,,	,,	7	,,
,,	35°	,,	30°	,,	,,	,,	15	,,
,,	30°	,,	25°	,,	,,	,,	19	,,
,,	25°	,,	20° there is only one observation of rain.					

NOVEMBER.

Pressure.—The highest readings on the chart, lie between 20° and 40° S., and west of the meridian of 92° W. Between this district and the coast, pressure is somewhat lower. The isobar of 30·1 in., appears to overlap the coast between the parallels of 30° and 40°. Its southern limit is in 42° S.

The only other isobar which is given is that of 29·3, lying nearly in latitude 55° S. In the high latitudes, the readings are very irregular, and as, near the parallel of 40°, the difference in barometrical readings amounts to 0·8 in. in about 700 miles, the westerly winds reach an average force of 7 or 8.

Wind.—The Trade wind extends to 30° S., and the southerly winds extend down the coast for 15° from that parallel.

The westerly winds in higher latitudes appear to have northing in them, except to the eastward of Patagonia, where they show, on the contrary, southing.

Temperature.—The isotherm of 65° has advanced slightly to the southward of its position in October, and passes from 28° S. 95° W., to the coast in 15° S. 75° W.

All the other isotherms lie nearly along the parallels of latitude, but for the most part bend to the northward along the coast. That of 40° is in 57° S.

Sea Temperature.—This agrees on the whole very fairly with that of the air.

Rain.—Between 60° and 55° rain has been noted once in every 2 observations.

,,	55°	,,	50°	,,	,,	3	,,
,,	50°	,,	35°	,,	,,	5	,,
,,	35°	,,	30°	,,	,,	10	,,
,,	30°	,,	20°	,,	,,	6	,,
,,	20°	,,	15°	,,	,,	18	,,
,,	15°	,,	10°	,,	,,	14	,,

North of 10° there are no records of rain.

DECEMBER.

Pressure.—The highest pressure is now in 35° S., and 95° W.

The isobar of 30·1 embraces a large area, from 17° to 44° S., and nearly up to the coast. It will be seen that on the chart the figures at the northern extremity of this line are 30·0 instead of 30·1.

The isobar of 29·8 lies in 47° S.

    ,,   29·5   ,,   52° S. approximately.

    ,,   29·0 appears to coincide pretty nearly with the parallel of 60° S.

Wind.—The Trade wind appears to extend to 35° S., though there is considerable unsteadiness of direction in the area between 25° and 35° S., and 80° and 95° W. The southerly wind on the coast extends to 40° S. In this region the wind appears to draw slightly outward from the district of greatest pressure, instead of remaining parallel to the isobars.

The westerly winds are strong between 45° and 50° S., as well as to the south of Cape Horn, and easterly winds are not very uncommon in this latter locality, but it must not be forgotten that the total number of observations is large, so that the per-centage of easterly winds is but small.

Temperature.—The isotherm of 70° appears in 20° S. 100° W. and runs to the N.N.E.

    ,,   65° stretches from 34° S. 100 W. to 22° S. on the coast.

    ,,   60° takes a N.E. course from 45° S. 100° W. to 31° S. 70° W.

    ,,   55° runs parallel to the last.

The other isotherms are not deflected to the northward in the same way, and that of 40° only appears in the extreme south-west corner of the chart.

Sea Temperature.—This does not differ materially from that of the air.

Rain.—This is still most frequent in the south.

Between 60° and 55° S. it occurs once in every 2 observations.

,,	55°	,,	45°	,,	,,	4	,,
,,	45°	,,	40°	,,	,,	7	,,
,,	40°	,,	30°	,,	,,	16	,,
,,	30°	,,	25°	,,	,,	12	,,
,,	25°	,,	20°	,,	,,	22	,,
,,	20°	,,	10°	,,	,,	7	,,

GENERAL REMARKS ON THE CHARTS.

In considering these charts the reader will be struck by the small number of observations in some of the squares, but it was thought that as they were sifted into months, navigators would prefer their remaining so to their being thrown together and brought out in charts for three months, by which the number of observations in each square would be increased, but their distinctive value lessened. For instance, in the four squares south of Cape Horn, between 55° to 60° S. and 60° to 80° W., the proportion of winds with easting in them for the three months January, February, and March, taken together, is 15 per cent., but in February there are only 8, whilst in March there are 17 per cent.

Again, it will be seen that in these months most of the observations lying between 20° to 30° S. and 85° to 100° W. were taken in January, and none in March. Then again, March shows an isotherm of 80°, whilst January only shows that of 70°. Similar important differences may be traced throughout the charts. It is always easy for the navigator to combine the data of any two or three charts, but impossible for him to sift them if they are once combined.

Instead of the materials having been sifted too nicely, captains who beat to the westward round Cape Horn still ask a very practical question which our charts do not answer, viz., Is the easterly wind which exists in the five degree squares south of Cape Horn evenly spread over those squares, or is it more common to the southward than close up to the land ?

To answer this question it would be necessary to work up the data of these few squares into one degree squares. It is, however, instructive to notice that the amount of easterly winds shown in high latitudes is greater in June than in January. In the above-named four squares between 55° to 60° S. and 60° to 80° W. there are 22 per cent. of winds with easting in them in June, whilst between 80° and 90° W., in the same month, they amount to 67 per cent., no west wind being recorded in 33 observations. Now according to Buys Ballot's law this indicates a higher pressure towards the pole in winter, similar to what is found to exist to the north of Iceland in the Atlantic. But whether this higher pressure over South Shetland than over the sea to the north of it prevails all the year round is still a disputed question ; if it does, a corresponding prevalence of easterly wind may be expected to exist near that land, while westerly winds are blowing near Cape Horn.

It is the opinion of some experienced men that this is the case, but the navigator must consider the risk of more ice and longer nights at certain seasons when tempted to go south in search of easterly winds.

The Admiralty pilot charts allude to the idea that better passages to the westward are made by going to lat. 60° S. in July, August, and September.

Although many parts of these charts suffer materially from want of data, there is such a general agreement amongst them that a cursory inspection of the isobars, isotherms, and wind arrows gives a good idea of the relation between pressure, temperature, and wind. The divergence of the isobars in about 40° to 45° S., part running to the north, whilst others run East or even S.E. (see the chart for March), with a corresponding divergence of the wind, is very instructive.

The way in which an area of high pressure exists over the sea somewhere between 22° and 42° S., but does not reach to the land, is remarkable. In connection with this area of high pressure it is an interesting fact that the temperature is considerably higher over the district where it is observed than it is near the coast. The highest barometer recorded between 50° and 60° S. was 30·603 in June, the lowest 27·662 in April, and these are probably the extremes of pressure for that latitude.

When considering the winds of corresponding latitudes in the North Atlantic, the navigator will see how this turning of the air, part towards the equator, and part towards the pole, corresponds with what is experienced to the westward of Portugal. *There*, in about 40° N., and on the western side of a great ocean, the wind seems to split into northerly along the coast (where many seamen say it is northerly for 10 months in the year), and S.W., or southerly, further north. This splitting of the air is also accompanied by an area of high pressure to the westward over the sea, and a lower pressure to the northward. (See Plate III. of the Barometer Manual, published by this office.)

It will be noticed that the wind seems to draw round the coast of Patagonia; for instance, the wind arrows indicate a more northerly direction on its west coast, more westerly on its south, and more southerly on its east coast. The swell of this latter wind seems to overrun it to a great distance; for instance, outward bound ships to India frequently experience a high S.W. swell at the southern verge of the S.E. trades in the Atlantic, though they rarely get the wind from that quarter. In a similar way at the northern verge of the N.E. trades, ships often get a high N.W. swell without its wind ; this swell probably coming from the strong N.W. winds which prevail in high latitudes on the western side of the Atlantic at certain seasons.

Between 50° and 60° S., both the air and sea are generally cooler to the eastward than to the westward of 75° W., the above-named tendency in the direction of the wind, and the well-known current to the N.E., as shown on the Admiralty pilot charts, have no doubt their influence in causing this result.

It need hardly be remarked that a larger number of careful observations is needed, especially in the central parts of the South Pacific. The Admiralty wind charts for this ocean now coming out, indicate that a good deal of northerly wind frequently exists where the S.E. trades might be expected, and from our general knowledge of the working of Buys Ballot's law, it is clear that if a space be found to the westward of the areas of high pressure given in these charts, where the pressure is lower, a northerly wind may be expected to blow there. This shows the great value of careful barometer readings in fine weather, as a sufficient number will indicate the direction of the prevailing wind.

The following table will enable the reader to compare the temperatures found within the area of high barometrical pressure and those observed on the coast.

Month.	Temperature in latitude 40° S.		Temperature in latitude 20° S.	
	Long. 100° W.	On the Coast.	Long. 100° W.	On the Coast.
January - -	65	58	73	69
February - -	62	60	—	70
March - -	62	58	75	70
April - -	60	57	72	66
May - -	50*	55	69	65
June - -	55	50	—	65
July - -	52	49	71	62
August - -	47†	50	65	61
September -	55	50	65	62
October - -	54	51	65	61
November -	55	53	—	64
December -	62	57	70	66
Mean - -	56·6	54·0	69·4	65·1

The question arises, what produces this difference between the littoral and oceanic climate ? The cause has hitherto been assigned to the cold water of the oceanic current discovered by Humboldt, which is said to exist off the coasts of Chili and Peru, and to transport the waters of latitude 50° into the equatorial westerly drift. The observations of the temperature of the surface water contained in these charts do not appear to give much support to this hypothesis, since the sea here is almost always warmer than the air. It would seem, therefore, that the air must cool the sea, rather than that the sea cools the air. The current may be an auxiliary cause, though we have not the means before us to investigate the matter ; but the principal cause seems to be the prevalent southerly wind. From lat. 40° S., sometimes from 45° S., or much further, a southerly wind generally blows along the coast, and eventually turns into the S.E. Trade, after passing the area of highest pressure. This wind only fails in July and August, which are precisely the months in which the isotherms of the middle latitudes are the least deflected northward. It seems therefore probable that the general deflection of the isotherms

---

* Very few observations give this result, in 95° W. the temperature is 60°.

† Very few observations in August, in 90° W. 25 observations give 54°.

C

northward, or in other words, the cool temperature off the coast, is brought about by this southerly wind which transfers the air of the temperate zone to the tropics.

The extension from the coast of this southerly wind is given approximately below:

January	to 85° W.	in 40° S.,	to 85° W.	in 30° S.	
February	to 80° ,,	,,	to 85° ,,	,,	
March	to 85° ,,	,,	to 80° ,,	25° S.	
April	to 85° ,,	45° S.,	to 82° ,,	,,	
May	to 85° ,,	,,	to 80° ,,	20° S.	
June	to 80° ,,	40° S.,	to 80° ,,	25° S.	
July	interrupted.				
August	interrupted.				
September	to 82° W.	in 40° S.,	to 82° ,,	30° S.	
October	to 80° ,,	,,	to 85° ,,	25° S.	
November	to 80° ,,	45° S.,	to 80° ,,	30° S.	
December	to 80° ,,	40° S.,	to 85° ,,	25° S.	

The southerly wind begins to fail in June, during which month, as well as in July and August, the atmospheric pressure off the coast is nearly similar to that over the ocean to the westward.

From March to August the Trade wind extends only to 25° S., during the rest of the year it reaches 30° S.

The arrows drawn upon the charts represent the prevalent winds. Generally they represent the current of air as flowing between the isobars conformably to the law of wind in relation to pressure in the southern hemisphere. Where the isobars are closest the winds are strongest, and where they diverge from each other light breezes are usually reported. Patches of high pressure seem to be frequently accompanied by light airs and calms. An irregular course of the isobars is generally indicative of atmospheric disturbance, or of variable winds.

Between 50° and 60° S., and 50° to 70° W., kelp or other sea-weed is frequently reported.

REMARKS ON WEATHER NOTATIONS.

Since preparing the wind charts for publication it has been decided that other weather data should be given. Here also the method of squares has been thought best, and a key diagram is given for each month, which render but few introductory remarks requisite.

It will be seen that the state of the weather has been indicated according to Beaufort's notation.

b Blue Sky.	g Gloomy.	o Overcast.
c Clouds (detached).	h Hail.	p Passing Showers.
d Drizzling Rain.	l Lightning.	q Squally.
f Foggy.	m Misty (hazy).	r Rain.

s Snow.
t Thunder.
u Ugly (threatening) appearance of Weather.

v Visibility. Objects at a distance unusually visible.
w Wet (Dew).

For convenience of reference the few remarks which are thought necessary have been printed on each chart.

Only 30 notations of lightning have been made ; these were well distributed between the parallels of 25° S. and 60° S. in the different months, excepting in January, in which month no lightning was recorded. In the Trade wind region from 25° S. to the equator, there was no lightning recorded. The notations of lightning are about four per cent. of the total number of weather observations in which it occurred.

The reader must bear in mind that the " No. of Weather Observations " in the upper left-hand corner of each square is the number of eight-hourly periods for which the weather was noted, and that several of the facts recorded may have existed during one of these periods, for instance, m, b, c, and q may have been entered for one, so that the sum of all entries will, in general, much exceed the number of eight-hourly periods. He must also remember that to get the relative proportion of certain weather in different squares, the number of observations in the upper left-hand corner of each square must be considered. For instance, in the January diagram the proportion of mist is 10 to 21, or nearly 50 per cent. near Valparaiso, whilst it is 17 to 84, or only 20 per cent., to the S.W. of Cape Horn. Now, if the number of observations be neglected, 17 gives the idea of more mist than 10. The same remark is applicable to the proportion of certain winds in different squares.

WEATHER OBSERVATIONS.

JANUARY.

KEY DIAGRAM.

	n m f	No. of Weather Observations.	m Mist or Haze.	f Fog.	STATE OF THE AIR.
	b o o				
	q s N°				
	b Blue Sky.		c Detached Clouds.	o Overcast.	STATE OF THE SKY.
	q Squalls.		s Mean Amount of Cloud.	N° No. of Observations of Cloud.	SQUALLS AND CLOUD OBSERVATIONS.

N.B.—The scale for the amount of cloud is the proportion of sky covered, 0 to 10; 0 = no clouds, 10 = entirely overcast.

Each number shows how often the weather indicated in the key diagram for its corresponding place has been recorded.

REMARKS.

Fogs are prevalent off the west coast of Patagonia, and mist off Valparaiso. It will be seen from the table at p. 34 that in this month the temperature at Santiago at an elevation of 1,782 feet is 4° higher than that at sea.

Squalls are often reported in high latitudes.

FEBRUARY.

REMARKS.

Mist is very common on the coast between 30 and 35 S.

Squalls are reported frequently at about 40° S., and also south of 40°, especially to the S.W. and S.E. of Cape Horn.

WEATHER OBSERVATIONS.

MARCH.

	100	95	90	85	80	75	70	65	60	55	50	
South 0	1 – – / – 6·0 1					n m f / b c o / q s N°						0 South
5												5
10	2 – – / – 3 – / – 4·5 3	2 – – / – 3 – / – 5·0 2										10
15		1 – – / – 1 3 / – 8·0 1		1 – – / – 1 – / – 7·0 1	10 – 2 / 1 4 3 / – 5·5 6							15
20			2 – – / – 2 – / – 6·5 2		3 – – / – 2 1 / – 7·3 3							20
25				10 – – / 2 3 5 / – 6·5 10	22 1 – / 3 8 9 / 2 6·9 19	6 – – / 1 1 3 / 2 7·2 6						25
30				12 – – / 3 7 2 / 2 4·6 11	19 1 – / 9 6 3 / 4 4·7 20	15 5 – / 3 5 7 / 3 7·1 14						30
35			15 1 – / 2 8 5 / 2 5·1 14	28 1 – / 7 8·6 18	3 – – / – 3 – / – 6·7 3	10 9 – / 1 6 8 / 1 6·1 9						35
40	6 3 – / – – 4 / – 10·0 4	12 1 – / 7 3 1 / 1 3·7 7	18 – – / 3 7 – / 8 8·1 16	6 – – / 1 4 2 / 3 6·8 6	37 2 – / 1 32 – / 5 4·7 26	22 – 1 / 0 11 4 / 2 3·7 17						40
45	3 1 – / – – 3 / – 10·0 3				15 – – / 1 6 7 / 5 4·3 5	34 – 1 / 1 13 16 / 7 7·6 21						45
50			3 – – / – 3 – / 3 5·7 3		8 – – / 4 5·0 1	36 6 2 / – 13 21 / 8 8·8 25	6 – – / 1 1 4 / – 7·0 4					50
55					43 2 – / 4 16 20 / 11 6·0 21	43 1 – / 1 20 22 / 14 7·8 32	40 1 – / – 20 20 / 10 8·9 28	27 1 – / 5 17 5 / 7 5·0 22	28 3 – / 2 15 11 / 10 6·8 36	17 5 1 / 1 9 4 / 1 6·6 17	13 6 2 / 1 9 4 / 4 5·9 7	55
60	22 5 2 / – 11 7 / 6 7·8 16	10 – – / – 5 5 / 5 7·7 6	13 2 – / – 1 12 / 1 10·0 7	16 3 – / – 6 10 / 6 8·2 13	60 2 – / 1 34 25 / 26 7·2 52	61 8 8 / 1 31 23 / 18 7·6 56	63 11 1 / 4 29 27 / 13 7·1 57	48 6 5 / 3 19 20 / 13 7·2 37	7 3 1 / 1 5 – / 2 5·7 6	4 3 – / – 3 – / – 6·0 3		60
West 100		95		90		85		80		75		70		65		60		55		50 West

Remarks. Mist is still reported near Valparaiso, and also near the Falkland Islands.

Squalls are not uncommon generally.

APRIL.

	100	95	90	85	80	75	70	65	60	55	50	
South 0						n m f / b c o / q s N°						0 South
5												5
10												10
15				5 – – / – 3 2 / 2 7·4 5	10 – – / 5 3 2 / – 5·0 8							15
20	1 – – / – 1 – / – 4·0 1			– – – / – 1 1 / – 6·7 3	3 – – / – 1 1 / 2 6·3 3	5 – 6 / – 7·0 8	11 – – / 1 3 7 / 1 9·0 9					20
25	2 – – / – 2 – / – 3·5 2	4 – – / 2 3 – / – 4·0 4		– – – / – 5·0 6	3 – – / 1 6·3 3	18 1 – / 3 9·4 10	37 6 – / 1 7 18 / – 7·9 22					25
30	6 1 – / 1 3 – / – 4·3 6		18 2 – / 2 10 4 / 1 5·3 18	3 – – / – 3·8 12	19 1 – / 11 6 1 / 2 3·5 17	9 – – / 4 4 – / 1 4·1 9	36 8 – / 3 15 17 / – 6·1 30					30
35	6 – – / 1 4 1 / 1 4·7 6		10 1 – / 6 3 – / 1 3·6 10	5 – – / 2 – 3 / – 4·4 14	23 1 – / 7 11 5 / 11 5·2 22	6 – – / 1 5 – / – 3·0 6	12 1 2 / 3 1 4 / 4 4·7 7					35
40			19 – – / 3 8·7 13	20 0 0 / 2 9 7 / 1 8·6 19	16 4 – / 2 11 4 / 9 5·5 15	13 6 – / 7 4 1 / – 3·7 13						40
45			9 – – / 1 – – / – 8·7 7	18 1 – / 2 10 5 / 6 7·7 15	22 4 – / 1 12 9 / 7 7·2 22							45
50				27 6 – / 3 13 7 / 7 0·1 17	23 5 – / 2 8 10 / 13 6·7 17							50
55	9 – – / – 3 6 / 1 7·6 9	3 – – / – 9·0 3		27 – – / 2 12 13 / 7 7·3 25	19 1 – / – 11 7 / 9 6·9 17	3 – – / – 2 1 / 1 8·7 3	1 – – / – 10·0 1	9 – – / 1 2 4·6 5	21 – 1 / 1 6·7 20	27 2 – / – 4 10 / 8 9·0 31		55
60	6 – 3 / 1 6·5 6	6 3 – / 1 2 – / 2 7·5 4	19 2 3 / 2 11 18 / 5 8·4 18	30 1 1 / 1 11 13 / 7 8·3 25	16 1 1 / 2 14 11 / 12 7·7 25	37 1 – / 2 14 13 / 18 6·7 25	58 6 5 / 20 8·9 42	35 1 6 / 3 4 22 / 11 8·0 26	15 1 – / 1 6 8 / 7 7·6 13			60
West 100		95		90		85		80		75		70		65		60		55		50 West

Remarks. Fog is not so often recorded. Mist is rather common along the coast, from 20° S. Squalls are rather frequent south of 30° S. In high latitudes they are very abundant, and also in 30°–35° S., 80°–85° W. In April the temperature of Santiago falls below that of the air of the sea near the coast. This fact may have to do with the decrease of mist.

MAY.

| 100 | 95 | 90 | 85 | 80 | 75 | 70 | KEY DIAGRAM. | 60 |

KEY DIAGRAM.

n No. of Weather Observations.	m Mist or Haze.	f Fog.	STATE OF THE AIR.
b Blue Sky.	c Detached Clouds.	o Overcast.	STATE OF SKY.
q Squalls.	a Mean Amount of Cloud.	N° No. of Observations of Cloud.	SQUALLS AND CLOUD OBSERVATIONS.

N.B.—The scale for the amount of cloud is the proportion of sky covered, 0 to 10; 0 = no clouds; 10 = entirely overcast.

Each number shows how often the weather indicated in the key diagram for its corresponding place has been recorded.

REMARKS.

Squalls are frequent between 40° and 60° S. The amount of cloud is great over the whole district.

JUNE.

| 100 | 95 | 90 | 85 | 80 | 75 | 70 | 65 | 60 | 55 | 50 |

REMARKS.

Mist is rather common on the coast from 10° to 15° S, but elsewhere weather is clearer than it was. Squalls are generally frequent south of 35° S.

JULY.

REMARKS.

From this month until December the air near Valparaiso is tolerably free from mist. Squalls are commonly reported. The amount of cloud is high and nearly uniform.

AUGUST.

REMARKS.

The weather resembles that of July.

It is very squally to the S.W. and W. of Cape Horn.

SEPTEMBER.

						KEY DIAGRAM.			
					n m f	n. No. of Weather Observations.	m Mist or Haze.	f Fog.	STATE OF THE AIR.
					b c o				
					q a N°	b Blue Sky.	c Detached Clouds.	o Overcast.	STATE OF SKY.
						q Squalls.	a Mean Amount of Cloud.	N° No. of Observations of Cloud.	SQUALLS AND CLOUD OBSERVATIONS.

N.B.—The scale for the amount of cloud is the proportion of sky covered, 0 to 10; 0=no clouds; 10 entirely overcast.
Each number shows how often the weather indicated in the key diagram for its corresponding place has been recorded.

REMARKS.
Squalls are frequent, especially between 15° and 35° S., and in high latitudes.

OCTOBER.

					n m f	
					b c o	
					q a N°	

REMARKS.
Mist is reported on the coast between 10° and 15° S.
Squalls are less frequent between the tropics, but are very common in the south, especially near Cape Horn.

NOVEMBER.

	100	95	90	85	80	75	70	65	60	55	50	
South 0					6 – – 1 5 – – 3'8 5		n m f b c o q a N°					0 South
5					6 3 – 1 – – – 3'8 6							5
10					13 2 – 1 11 1 – 4'3 12	1 – – – 1 – – 7'0 1						10
15				4 – – – 4 – – 7'7 3	5 – 3 – 3 3 – 8'7 3	8 2 – – 4 4 – 7'5 8	5 – – – 9'5 5					15
20			6 – – – 6 – – 5'8 4	4 – – – 1 3 – 8'0 5	17 – – – 11 5 – 6'1 17	1 – – – 9'0 1						20
25		19 – – 4 14 – 2 4'7 15	15 – – 3 8 4 3 6'6 14	24 – – – 15 9 – 6'7 16	14 2 – – 9 4 – 5'6 7							25
30		15 – – 9 9 7 – 5'4 25	10 – – 4 5 – 2 3'9 9	11 1 – 1 7 2 – 6'4 11	16 1 – – 9 7 – 5'4 10							30
35		7 – – – 6 1 – 6'4 7	17 1 – 2 13 10 1 7'0 15	10 1 – – 5 5 – 7'8 10	15 – – 1 5 9 – 7'2 11		REMARKS. Fog and mist are not uncommon in high latitudes. Squalls are rare north of 40° S., but are common to the southward, especially near the land.						35
40		2 – – 1 – – 1 2'0 1	11 2 4 1 15 10 2 7'9 15	21 1 – 1 12 8 4 6'5 19	8 – – 1 6 1 1 6'8 4							40
45			17 4 7 1 1 5 – 8'9 11	28 2 – 3 17 6 2 5'7 23	21 – 1 1 7 10 3 5'5 13							45
50				40 4 9 3 15 11 3 6'8 27	63 3 4 3 36 27 25 6'0 34	12 – – 3 8 1 9 4'6 11	46 3 – 9 19 14 15 5'7 38	63 6 – 6 32 22 13 6'1 63	48 – – 7 17 16 11 6'2 45	20 1 – 7 6 7 3 5'2 20		50
55	3 – – – – 3 1 10'0 3	6 2 1 – 1 3 2 9'0 6	6 3 – – 3 3 – 9'0 6	17 3 1 3 10 5 1 6'6 16	55 4 11 4 16 23 11 6'7 35	54 5 4 3 22 25 7 7'8 50	103 3 6 4 48 43 23 7'7 103	64 5 3 4 35 21 10 6'9 60	2 – 1 – – 1 – 10'0 1	1 – – – 1 – 1 7'0 1		60
West	100	85	90	95	80	75	70	63	60	55	50 West	

DECEMBER.

	100	95	90	85	80	75	70	65	60	55	50	
South 0							u m f b c o q a N°					0 South
5												5
10	7 – – 3 2 3 – 3'3 4											10
15	15 – – 5 10 3 – 5'7 15	6 – – – 6 – – 6'8 4		6 – – 1 4 1 – 6'2 6	9 – 1 3 5 – – 4'5 12			~				15
20	7 – – – 4 3 1 7'9 7	25 – – 2 16 7 1 6'4 21	16 – – 2 9 5 1 5'9 14	9 – – – 6 3 3 7'5 12	6 – – 3 – 3 – 5'9 9							20
25	3 – – – 3 – – 5'0 3	6 – – – 3 3 – 8'0 7	15 1 – 3 7 10 3 7'9 10	15 – – 6 5 4 – 5'0 21	18 – – – 4 14 – 9'2 18							25
30	3 – – – 3 – – 5'7 3	14 – – 5 7 14 5 7'0 14	6 – – – 3 3 – 5'9 21	25 – – 9 13 3 2 3'8 22	33 2 – 2 10 20 2 8'1 29	4 – – 2 1 – – 3'7 3		REMARKS. Fog and mist extend up to 30° S., but are rarely recorded north of that latitude.				30
35	15 – – 5 9 1 1 3'5 24	24 – – 7 13 4 4 4'7 25	12 – – – 6 4 4 5'6 8	38 – – 5 22 8 4 8'0 35	12 3 – 5 7 – – 3'9 12	11 2 – 3 4 3 6 6'0 10		Squalls are still very common in the high southern latitudes.				35
40	4 – 1 – 8'0 2	18 1 – 4 11 2 – 5'3 15	18 – – 3 9 4 3 5'6 15	24 2 – 6 10 8 4 5'6 24	36 3 6 6 16 11 – 6'5 35	11 2 1 – 7 3 – 6'5 11						40
45	9 – – 1 4 4 – 7'8 9	14 1 – 4 6 8 – 6'5 11	18 5 – 3 7 7 3 7'7 17	32 2 1 2 19 9 3 6'5 33	30 8 – – 11 7 2 7'7 20							45
50		6 1 3 – 9'8 6	23 1 3 5 9 14 5 7'8 20	36 13 3 1 15 12 9 8'1 37	51 6 – – 11 17 8 8'4 30							50
55		3 – – – 3 – 3 5'3 3	53 7 – 1 25 21 11 7'5 51	65 6 2 – 22 40 11 8'6 62			21 4 6 – 14 3 1 7'7 21	26 4 – 4 16 – – 6'1 18	14 2 1 – 9 4 1 6'8 11	15 5 – 1 6 6 – 7'0 12		55
60	9 1 – – 3 – – 6'3 3	3 – – 1 – 2 – 8'0 3	23 – – 1 12 9 3 6'6 21	85 3 2 – 50 32 24 6'1 83		101 3 4 3 45 52 23 7'7 93	84 18 4 – 34 37 29 9'1 59	43 10 3 1 19 13 11 8'7 34	8 1 1 1 4 2 1 8'0 6			60
West	100	95	90	85	80	75	70	65	60	55	50 West	

27687.

D

ADDITIONAL DATA.

In addition to the information which is contained in the charts, the Office is in possession of observations made at certain stations situated within the area to which the charts refer, and it has been considered that the publication of these materials would not be unacceptable. Their nature is as follows:

Table I. Tables of diurnal range of the barometer at Port Louis, East Falkland Island, and at Hermit Island, near Cape Horn, calculated by Captain Sir J. Clark Ross, R.N., F.R.S., from the observations made on board H.M.S. "Erebus" in 1842.

They exhibit a mean daily oscillation of the barometer of 0·016 in. between 9.30 a.m. and 3.30 p.m. This value agrees very closely with the theoretical oscillation 0·021 in. computed from the following formula given by Professor James Forbes:—

$$x = -·015 + ·1193 \cos^2 \text{lat.}$$

Table II. Means of barometrical observations taken at Cape Pembroke Lighthouse, Falkland Islands, with a marine barometer, Kew pattern, belonging to the Meteorological Office.

Table III. Results of observations taken in the harbour of Valparaiso.

A. On board H.M.S. "Nereus." B. On board various other ships.

TABLE I.

MEAN PRESSURE of the ATMOSPHERE determined from HOURLY OBSERVATIONS of the BAROMETER, reduced to 32° and the sea-level, made on board H.M.S. "Erebus," Captain the late Sir JAMES CLARK ROSS, R.N., F.R.S.

Hours.	At Port Louis, East Falkland Island.						St. Martin's Cove.
	1842. April 6 to 30.	1842. May.	1842. June.	1842. July.	1842. August.	1842. Nov. 14 to Dec. 16.	1842. Sept. 20th to Nov. 6th.
1 a.m.	29·534	29·318	29·400	29·688	29·665	29·462	29·370
2 „	·535	·319	·394	·681	·661	·459	·367
3 „	·531	·313	·391	·684	·654	·457	·368
4 „	·528	·312	·390	·670	·649	·458	·367
5 „	·531	·310	·385	·655	·645	·464	·364
6 „	·537	·306	·384	·652	·641	·465	·366
7 „	·537	·308	·385	·651	·641	·467	·367
8 „	·539	·311	·384	·650	·643	·465	·363
9 „	·537	·316	·390	·652	·642	·457	·360
10 „	·534	·312	·394	·651	·642	·454	·360
11 „	·529	·313	·400	·651	·643	·453	·355
Noon	·525	·304	·397	·645	·637	·451	·348
1 p.m.	·511	·292	·353	·637	·629	·447	·347
2 „	·504	·286	·346	·632	·626	·446	·346
3 „	·503	·286	·381	·639	·628	·447	·343
4 „	·502	·290	·389	·645	·636	·449	·346
5 „	·504	·292	·396	·656	·644	·451	·348
6 „	·507	·295	·400	·665	·651	·454	·352
7 „	·510	·296	·406	·669	·656	·454	·362
8 „	·513	·299	·410	·670	·662	·457	·364
9 „	·514	·300	·413	·677	·667	·458	·365
10 „	·519	·300	·415	·681	·669	·462	·367
11 „	·525	·305	·414	·684	·666	·462	·369
Midnight	·527	·308	·414	·686	·669	·461	·369
Mean	29·522	29·304	29·393	29·661	29·648	29·457	29·360
No. of days	25	31	30	31	31	33	48

At Port Louis the ship's position was lat. 51° 33' S., long. 58° 7' W. At St. Martin's Cove, Hermit Island, the ship's position was lat. 55° 52' S., long. 67° 33' W.

TABLE II.

CAPE PEMBROKE LIGHTHOUSE, EAST FALKLAND ISLAND.

Latitude 51° 40' 42" S. Longitude 57° 42' 0" W.

MONTHLY AVERAGES of Barometrical Observations taken during the years 1859 to 1868 (not continuous). The results are obtained from four observations daily, viz., 4 a.m., 9 a.m., 3 p.m., and 8 p.m.

Observations corrected and reduced to 32° Fahrenheit at mean sea level and made with barometer verified at Kew.

Date.	January.	February.	March.	April.	May.	June.	July.	August.	September.	October.	November.	December.	Year.	Date.
1859	—	—	—	—	—	—	29·612	29·528	29·566	29·523	29·578	29·537	—	1859
1860	29·373	29·466	29·406	29·479	29·617	29·884	29·526	29·510	29·593	29·611	29·349	29·304	29·510	1860
1861	29·447	29·570	29·439	29·468	29·433	29·763	29·453	29·491	29·766	29·814	29·458	29·528	29·553	1861
1862	29·439	29·588	29·383	29·491	29·355	*29·483	29·579	29·403	29·770	29·660	29·471	29·431	29·506	1862
1863	29·252	29·531	†29·424	No observations.	29·793	29·438	29·259	29·455	29·630	29·648	29·596	29·240	29·479 11 months only.	1863
1864	29·593	29·537	29·446	29·605	29·434	29·626	29·574	29·478	29·696	29·691	29·565	29·425	29·555	1864
1865	29·528	29·752	29·714	29·507	29·403	29·715	29·428	29·543	‡29·531	No observations.			—	1865
1866	29·567	29·588	29·671	29·504	29·543	29·766	29·796	29·662	29·721	29·553	29·462	29·516	29·612	1866
1867	29·573	29·635	29·355	29·550	29·545	29·538	29·564	29·525	29·845	29·717	29·652	29·461	29·580	1867
1868	29·378	29·395	29·492	29·510	29·446	29·595	29·452	—	—	—	—	—	—	1868
Final average	29·461 9 years.	29·562 9 years.	29·481 9 years.	29·514 8 years.	29·507 9 years.	29·645 9 years.	29·514 10 years.	29·511 9 years.	29·680 9 years.	29·652 8 years.	29·516 8 years.	29·433 8 years.	29·541 From figs. to left.	Final average.

* Three days observations missing. † Seven days observations missing, from 25th to 31st inclusive. ‡ Two days observations missing.

TABLE III.

A.—Results of Meteorological Observations made at Valparaiso by the Officers of H.M. S. "Nereus."
Lat. 33° 1' 55" S. Long. 71° 40' 25" W.

Month. 1853–8.	Number of Years represented.	Barometer.		Temp. of Air.		Temp. of Evap.		Cloud.		Registering Thermometers.			
		Average at 32° and Sea Level.	No. of Observations.	Average.	No. of Observations.	Average.	No. of Observations.	Average Amount (0 to 10).	No. of Observations.	Mean Max. in Air.	No. of Observations.	Mean Min. in Air.	No. of Observations.
		At 9h. 30m. A.M.											
		Inches.											
January -	3	29·990	31	65·4	93	61·1	93	4·3	93	70·7	93	65·1	93
February -	4	—	—	63·4	113	59·7	113	4·6	113	68·5	113	62·9	113
March -	4	—	—	61·5	124	58·2	124	5·4	124	66·6	124	61·7	124
April -	4	—	—	59·8	120	57·4	90	4·9	120	64·0	120	59·9	120
May -	5	30·107	31	58·1	153	55·5	124	5·4	153	62·9	153	59·1	153
June -	4	30·104	30	56·9	120	53·7	90	5·6	120	61·5	120	57·5	120
July -	5	30·077	31	56·8	155	53·5	124	5·3	155	60·5	155	56·5	155
August -	5	30·139	62	56·8	155	53·4	124	5·5	155	61·0	155	56·7	155
September -	4	30·141	30	57·2	120	54·4	90	5·0	120	61·5	120	56·9	120
October -	5	30·139	31	58·9	154	55·3	124	4·6	154	63·6	154	58·4	154
November -	5	30·090	30	61·1	150	57·3	150	3·2	150	67·0	150	60·5	150
December -	5	30·022	31	63·2	155	59·2	155	3·3	155	68·7	155	62·7	155

(continued.)

Month.	Barometer.		Temp. of Air.		Temp. of Evap.		Cloud.	
	Average at 32° and Sea Level.	No. of Observations.	Average.	No. of Observations.	Average.	No. of Observations.	Average Amount (0 to 10).	No. of Observations.
	At 3h. 30m. P.M.							
	Inches.							
January - -	29·952	31	67·5	93	62·4	93	1·4	93
February - -	—	—	67·2	113	61·1	113	2·1	113
March - -	—	—	64·9	124	59·8	124	2·8	124
April - -	—	—	62·4	120	58·8	90	4·4	120
May - -	30·073	31	59·8	153	56·2	124	5·2	153
June - -	30·054	30	58·2	120	54·5	90	5·2	120
July - -	30·035	31	57·8	155	53·9	124	5·0	155
August - -	30·093	62	57·9	155	54·0	124	4·0	155
September - -	30·098	30	59·0	119	55·6	90	3·7	119
October - -	30·121	31	62·2	155	56·7	124	4·5	155
November - -	30·078	30	65·5	150	58·9	150	2·0	150
December - -	29·978	31	67·1	155	60·6	155	1·9	155

Months.	Total Observations of Wind.	9h. 30m. A.M. Observations of Wind, referred to 18 points. (Force 0 to 12.)*																	
		N.		N.N.E.		N.E.		E.N.E.		E.		E.S.E.		S.E.		S.S.E.		S.	
		O.	F.	O.	F.	O.	F.	O.	F.	O.	F.	O.	F.	O.	F.	O.	F.	O.	F.
January -	93	20	1·1	—	—	7	1·1	—	—	3	1·3	—	—	—	—	—	—	5	2·6
February -	113	20	1·1	3	1·0	14	1·0	—	—	3	1·0	—	—	1	1·0	—	—	3	2·7
March - -	124	33	1·4	3	1·0	13	1·2	—	—	6	1·0	—	—	1	1·0	—	—	3	2·0
April - -	120	40	1·4	3	1·0	4	1·0	1	1·0	2	1·0	1	1·0	3	1·0	1	1·0	8	1·8
May - -	153	36	1·5	5	1·4	17	1·1	1	1·0	3	1·0	1	1·0	3	1·3	1	1·0	9	1·4
June - -	130	24	2·7	11	1·5	9	1·6	1	1·0	4	1·7	—	—	3	1·3	—	—	2	1·0
July - -	155	34	2·1	1	4·0	23	1·8	2	1·0	15	1·5	—	—	12	1·4	1	2·0	8	1·3
August - -	155	53	2·0	8	1·8	8	1·1	1	1·0	7	1·4	—	—	1	1·0	2	4·0	16	2·0
September -	120	31	1·7	—	—	8	1·6	—	—	9	1·1	1	1·0	1	1·0	—	—	4	2·0
October -	154	24	1·5	20	1·7	19	1·3	1	1·0	13	1·4	—	—	2	2·0	—	—	27	2·1
November -	150	22	1·2	4	1·2	11	1·0	3	1·3	6	1·0	—	—	4	1·0	2	5·5	25	1·5
December -	155	39	1·9	2	1·5	12	1·0	1	1·0	6	1·2	—	—	1	1·0	—	—	10	2·4

(continued.)

Months.	9h. 30m. A.M. Observations of Wind, referred to 16 points. (Force 0 to 12.)														Variables.		No. of Calms.
	S.S.W.		S.W.		W.S.W.		W.		W.N.W.		N.W.		N.N.W.				
	O.	F.	O.	F.	O.	F.	O.	F.	O.	F.	O.	F.	O.	F.	O.	F.	
January -	2	3·5	4	2·0	—	—	6	2·0	—	—	13	1·2	1	1·0	—	—	32
February -	—	—	5	2·2	—	—	7	1·6	—	—	7	1·4	2	1·5	—	—	48
March - -	—	—	4	2·8	—	—	4	1·0	—	—	14	1·4	2	1·0	—	—	41
April - -	—	—	1	1·0	—	—	1	1·0	—	—	11	1·1	4	1·7	—	—	40
May - -	1	3·0	1	1·0	-	—	1	3·0	—	—	5	1·4	3	1·7	1	3·0	65
June - -	—	—	3	1·3	—	—	1	1·0	—	—	2	2·0	2	1·5	—	—	58
July - -	—	—	4	1·5	—	—	1	3·0	—	—	6	1·2	—	—	—	—	48
August - -	1	1·0	1	1·0	—	—	—	—	—	—	9	1·3	4	2·0	—	—	44
September -	2	2·0	7	1·7	1	2·0	3	1·7	—	—	20	1·4	3	2·0	—	—	30
October -	6	2·8	9	3·3	—	—	5	1·6	—	—	9	1·4	7	1·6	1	3·0	11
November -	6	3·7	9	1·7	—	—	5	2·0	1	1·0	10	1·4	5	1·4	1	1·0	36
December -	5	2·8	13	2·3	—	—	6	1·2	—	—	17	1·4	2	1·5	—	—	41

* O. Number of observations of wind.
F. Force.
The direction of the wind is not corrected for the variation of the compass.

Months.	Total Observations of Wind	2h. 30 m. P.M. Observations of Wind, referred to 16 Points. (Force 0 to 12.)*															
		N.		N.N.E.		N.E.		E.N.E.		E.		E.S.E.		S.E.		S.S.E.		S.
		O.	F.	O.	F.	O.	F.	O.	F.	O.	F.	O.	F.	O.	F.	O.	F.	O.	F.
January -	92	9	1.2	—	—	—	—	—	—	1	2.0	—	—	—	—	—	—	32	4.2
February -	113	2	1.0	1	1.0	3	1.0	1	1.0	3	1.3	—	—	3	1.0	—	—	45	4.0
March -	124	12	2.5	3	1.7	3	1.3	—	—	3	1.3	—	—	7	1.4	—	—	39	4.2
April -	120	14	2.1	1	1.0	4	1.2	—	—	5	1.4	—	—	6	1.2	—	—	31	3.0
May -	153	33	2.1	6	1.3	5	1.4	—	—	—	—	—	—	4	1.7	5	2.0	24	2.6
June -	120	25	3.4	4	1.2	4	2.0	1	2.0	2	1.0	1	1.0	5	1.8	3	2.0	21	2.7
July -	155	37	2.6	3	2.0	2	3.0	—	—	2	1.5	—	—	5	2.2	7	2.6	27	2.3
August -	155	33	2.7	—	—	8	1.8	1	2.0	1	1.0	1	5.0	3	3.7	7	2.6	39	2.8
September -	119	16	2.4	1	1.0	3	1.3	—	—	5	1.2	—	—	2	1.5	1	1.0	25	2.9
October -	155	6	1.7	5	2.2	4	1.5	2	2.0	6	1.7	—	—	2	2.5	—	—	46	3.5
November -	150	2	1.0	1	1.0	1	3.0	2	1.5	3	2.3	—	—	3	1.7	—	—	64	4.6
December -	155	9	2.0	—	—	3	1.3	—	—	3	1.3	—	—	1	2.0	—	—	53	4.1

(continued.)

Months.	2h. 30m. P.M. Observations of Wind, referred to 16 Points. (Force 0 to 12.)															Variables.		No. of Calms.
	S.S.W.		S.W.		W.S.W.		W.		W.N.W.		N.W.		N.N.W.				
	O.	F.	O.	F.	O.	F.	O.	F.	O.	F.	O.	F.	O.	F.	O.	F.	
January -	12	3.3	22	4.0	—	—	3	2.0	2	1.5	6	1.7	—	—	1	1.0	4
February -	15	3.9	16	2.8	3	1.3	8	1.5	—	—	4	1.7	—	—	1	3.0	8
March -	14	3.6	19	2.3	—	—.	4	2.8	1	2.0	6	1.3	1	1.0	—	—	35
April -	26	2.5	9	2.1	1	2.0	4	2.5	—	—	4	1.5	2	1.0	—	—	13
May -	12	2.4	12	2.3	2	1.5	2	1.5	—	—	7	2.0	1	2.0	2	1.0	38
June -	2	2.5	7	2.0	1	2.0	2	2.0	—	—	1	1.0	3	1.3	2	2.0	36
July -	8	2.1	13	2.1	1	2.0	3	1.3	—	—	5	1.4	1	3.0	—	—	41
August -	13	2.5	18	2.2	1	2.0	3	2.7	1	2.0	7	2.3	3	1.0	—	—	16
September -	14	2.9	16	2.9	5	2.6	1	1.5	—	—	8	1.5	—	—	—	—	21
October -	42	3.2	23	3.6	—	—	2	4.5	—	—	7	1.7	6	1.7	1	1.0	3
November -	29	3.9	27	3.6	—	—	1	2.0	1	1.0	5	2.4	3	2.7	—	—	8
December -	34	3.7	15	3.1	1	3.0	8	2.3	—	—	4	1.5	4	2.0	3	1.0	16

* O. Number of observations of wind.
F. Force.
The direction of the wind is not corrected for the variation of the compass.

B.—In VALPARAISO HARBOUR.

Months.	Barometer.		Temp. of Air.		Cloud.		Weather.								Temp. of Sea.	
	Average, at 32° and Sea Level.	No. of Observations.	Average.	No. of Observations.	Average Amount (0 to 10).	No. of Observations.	Total Obser- vations.	Sky.			Atmos- phere.		Rain- fall.		Average.	No. of Observations.
								b.	c.	o.	m.	f.	r. and h.	q.		
	Inches.		°												°	
January	—	—	·	—	—	—	—	—	·	—	—	—	—	—	·	—
February	—	—	—	—	—	—	—	—	—	—	—	—	—	—	—	—
March -	30·026	9	58·4	9	—	—	7	—	5	—	2	—	—	—	—	—
April -	30·084	14	61·4	14	1·0	2	14	11	3	—	—	—	—	—	—	—
May -	30·049	3	57·7	3	2·0	1	3	1	2	—	—	—	—	—	—	—
June -	—	—	—	—	—	—	—	—	—	—	—	—	—	—	—	—
July -	30·080	24	56·2	24	5·0	24	24	8	10	5	2	—	3	—	—	—
August	30·118	93	55·6	64	3·6	89	93	31	56	5	19	—	6	1	55·7	26
September	30·130	93	54·7	90	2·8	71	93	42	48	3	7	7	4	4	51·4	24
October	30·040	108	59·9	83	2·9	89	108	45	60	2	9	4	2	7	53·3	25
November	—	—	—	—	—	—	—	—	—	—	—	—	—	—	—	—
December	—	—	—	—	—	—	—	—	—	—	—	—	—	—	—	—

Months.	Total Observa- tions of Wind.	Observations of Wind, referred to 16 Points. (Force 0 to 12.)															
		N.		N.N.E.		N.E.		E.N.E.		E.		E.S.E.		S.E.		S.S.E.		S.
		O.	F.	O.	F.	O.	F.	O.	F.	O.	F.	O.	F.	O.	F.	O.	F.	O.	F.
January	—	—	—	—	—	—	—	—	—	—	—	—	—	—	—	—	—	—
February	—	—	—	—	—	—	—	—	—	—	—	—	—	—	—	—	—	—
March -	9	—	—	—	—	—	—	—	—	—	—	—	—	—	—	—	1	2·0
April -	14	—	—	—	—	—	—	—	—	—	—	—	—	—	—	—	1	1·0
May -	3	—	—	—	—	—	—	—	—	—	—	1	1·0	—	—	—	—	—
June -	—	—	—	—	—	—	—	—	—	—	—	—	—	—	—	—	—	—
July -	24	5	4·2	—	—	—	—	—	—	2	2·5	—	—	2	3·0	—	—	5	2·6
August -	93	9	2·1	4	2·0	—	—	—	—	1	1·0	—	—	—	—	2	3·5	2	1·5
September	93	1	1·0	1	1·0	—	—	—	—	—	—	1	1·0	1	3·0	4	2·5	10	2·7
October	108	3	1·7	2	1·0	—	—	1	1·0	1	1·0	—	—	3	2·3	2	2·5	9	2·4
November	—	—	—	—	—	—	—	—	—	—	—	—	—	—	—	—	—	—
December	—	—	—	—	—	—	—	—	—	—	—	—	—	—	—	—	—	—

(continued.)

Months.	Observations of Wind, referred to 16 Points. (Force 0 to 12.)														Variables.		No. of Calms.
	S.S.W.		S.W.		W.S.W.		W.		W.N.W.		N.W.		N.N.W.				
	O.	F.	O.	F.	O.	F.	O.	F.	O.	F.	O.	F.	O.	F.	O.	F.	
January	—	—	—	—	—	—	—	—	·	—	—	—	—	—	—	—	—
February	—	—	—	—	—	—	—	—	—	—	—	—	—	—	—	—	—
March -	—	—	—	—	—	—	—	—	—	—	—	—	—	—	—	—	8
April -	2	1·5	3	3·3	1	2·0	—	—	—	—	—	—	—	—	—	—	7
May -	1	4·0	—	—	—	—	1	1·0	—	—	—	—	—	—	—	—	—
June -	—	—	—	—	—	—	—	—	—	—	—	—	—	—	—	—	—
July -	—	—	—	—	—	—	—	—	—	—	1	3·0	—	—	3	—	6
August -	5	1·4	5	2·2	1	1·0	—	—	1	2·0	5	1·8	1	2·0	5	1·2	52
September	3	2·3	8	2·1	1	3·0	3	1·7	—	—	2	1·0	1	4·0	11	1·0	46
October	10	2·9	10	2·1	1	3·0	1	1·0	1	2·0	2	1·0	4	1·5	4	1·3	54
November	—	—	—	—	—	—	—	—	—	—	—	—	—	—	—	—	—
December	—	—	—	—	—	—	—	—	—	—	—	—	—	—	—	—	—

APPENDIX.

The following additional information respecting the climate of South America is given, on the authority of Dr. J. Hann, in the *Journal of the Austrian Meteorological Society*, Vols. V. and VI., the data being here converted to English measures:—

PUNTA ARENAS, lat. 53° 12′ S., long. 70° 56′ W.

Months.	Temperature.			Rainfall.	
	Mean.	Mean		Mean Amount.	No. of Days.
		Maximum.	Minimum.		
	°	°	°	Inches.	
January	51·4	71·4	44·2	1·429	13·1
February	50·4	65·7	42·8	2·012	13·7
March	47·1	63·1	39·4	1·941	16·1
April	42·1	55·9	31·6	2·654	14·2
May	38·8	51·1	30·2	1·756	11·8
June	35·8	45·0	24·8	2·153	11·6
July	34·5	46·2	22·8	2·736	14·4
August	36·0	50·2	26·1	1·862	11·6
September	39·7	54·0	30·7	1·067	11·3
October	43·9	58·5	33·6	1·350	10·7
November	46·8	60·1	39·7	1·303	14·3
December	49·6	65·3	43·3	1·433	16·0
Year	43·0	74·3	17·6	21·696	158·8
No. of years	6	5	5	6	6

Frequency of Winds in Per-centages.

—	N.	N.E.	E.	S.E.	S.	S.W.	W.	N.W.
Summer	10	5	4	1	7	13	41	19
Autumn	16	13	4	2	5	16	29	16
Winter	19	12	7	0	3	12	31	16
Spring	12	7	3	1	7	14	37	20
January	9	4	4	1	7	15	41	20
July	21	13	10	1	4	13	26	12

The hours of observation were 8 a.m., noon, and 8 p.m., but a correction has been made by Dr. Hann to obtain the mean day temperature.

The maximum and minimum temperatures for the year are absolute extremes.

Puerto Montt, lat. 41° 30′ S., long. 72° 52′ W., 32·3 feet above the sea.

Months.	Barometer.	Temperature.			Rainfall.	
		Mean.	Mean		Mean Amount.	Days.
			Maximum.	Minimum.		
	Inches.	°	°	°	Inches.	
January - - -	29·934	59·7	76·3	49·4	3·543	8·5
February - -	29·926	59·4	75·9	48·4	4·303	8·6
March - - -	29·945	56·7	73·2	46·6	7·488	10·7
April - - -	30·012	52·5	68·9	40·8	11·874	14·2
May - - -	30·012	48·6	62·8	35·8	14·988	17·3
June - - -	30·001	45·9	56·7	32·7	15·067	18·6
July - - -	29·843	44·8	56·5	33·8	12·453	17·7
August - -	29·981	45·5	57·4	35·6	9·153	15·6
September -	30·060	47·3	59·2	36·7	6·988	13·7
October - -	30·075	50·4	68·2	41·2	6·142	12·3
November -	30·004	54·1	66·2	44·1	5·571	11·1
December	30·036	57·7	79·7	48·4	4·485	9·7
Year - - -	29·989	51·8	81·3*	30·9*	102·06	158·
No. of years - -	1½	6	3	3	6	6

Frequency of Winds in Per-centages.

	N.	N.E.	E.	S.E.	S.	S.W.	W.	N.W.
Summer - - -	30	10	1	18	39	1	1	10
Autumn - - -	50	2	0	11	15	4	3	14
Winter - - -	67	2	1	5	8	3	3	12
Spring - - -	42	2	1	15	22	3	1	15

The hours of observation were 8 a.m, 2 p.m., and 8 p.m., and a correction has been applied to obtain the mean temperature.

* Mean yearly range 1859–64.

VALDIVIA, lat. 39° 49′ S., long. 73° 13′ W.; 12 years, 1853–64.

| Months. | Temperature. | | | Rainfall. | |
| | Mean. | Mean· | | Mean Amount. | Days. |
		Maximum.	Minimum.		
January	62·8	87·8	44·4	2·559	5·8
February	61·3	83·1	42·8	3·937	7·2
March	57·9	78·3	40·3	5·984	8·5
April	53·8	67·3	37·2	10·945	11·7
May	49·8	61·7	33·8	15·315	14·8
June	46·8	57·7	31·8	22·165	18·4
July	45·1	56·1	31·8	16·024	15·4
August	46·6	59·2	30·2	12·165	14·4
September	48·6	66·2	31·8	6·496	10·6
October	52·9	73·8	36·0	4·567	7·8
November	56·3	77·2	40·1	5·709	9·4
December	61·0	86·0	43·9	3·110	7·4
Year	53·6	—	—	108·976	131·4

The hours of observation were 6 a.m., 2 p.m., 10 p.m., and a correction has been applied to obtain the mean temperature.

SANTIAGO, lat. 33° 26·4′ S., long. 70° 37′ W., 543 metres = 1,782 feet above the sea.

N.B.—In the original paper, Vol. V., p. 441, the height is given as 569 metres, but in Vol. VI., p. 109, this is amended to the figure adopted in the text.

| Months. | Barometer, mean, at 32° F. | Barometer reduced to Sea Level. | Temperature. | | | Rainfall. |
| | | | Mean. | Mean | | Amount. |
				Maximum.	Minimum.	
	Inches.					
January	28·174	30·015	66·0	85·6	52·9	0·000
February	28·174	·015	65·1	86·4	52·2	0·051
March	28·193	·046	61·5	82·0	46·0	0·098
April	28·229	·100	55·6	72·7	40·5	0·539
May	28·252	·142	49·3	66·6	37·4	2·031
June	28·260	·160	46·0	63·7	35·6	3·938
July	28·272	·171	45·1	61·3	33·4	4·165
August	28·296	·197	46·9	63·9	34·3	2·772
September	28·272	·161	51·3	72·3	37·6	1·646
October	28·260	·136	55·4	78·4	41·9	0·708
November	28·225	·082	60·4	79·5	46·8	0·590
December	28·186	·028	64·8	84·9	52·5	0·252
Year	28·233	30·104	55·6	87·8*	31·6*	16·790
No. of Years	6½	—	8½	7½	8	11

* Absolute maximum and minimum.

Frequency of Winds in Per-centages.

	N.	N.E.	E.	S.E.	S.	S.W.	W.	N.W.	Calms.†
Summer - - -	5·5	12·5	4·5	5·5	8	44·5	12·5	7·5	16·3
Autumn - - -	7	21	8·5	8	11·5	27	6·5	9·5	16·4
Winter - - -	7	28	8	10	10·5	16	5·0	16·5	25·0
Spring - - -	3·5	14	4	8·5	10	38	11·5	11·0	16·4

† The calms are expressed in per-centages of the total number of observations.

COPIAPO.

Latitude, 27° 22′ 30″ S.   Longitude, 70° 23′ W.   Height above the sea, 1,296 feet.

Two Years, 1864 and 1867.	Barometer.			Temperature.				Cloud.
	Mean at 32° F.	Daily Range.	Absolute Range.	True Mean.	Mean of 9 a.m. and 9 p.m.	Mean Daily Maximum.	Mean Daily Minimum.	Mean Amount (0 to 10).
	Inches.	Inch.	Inch.	°	°	°	°	
January - - -	28·651	·051	·252	68·7	67·1	83·3	57·6	1·4
February - - -	28·629	·043	·213	68·7	66·6	82·9	58·8	1·5
March - - -	28·641	·055	·264	66·6	64·4	80·6	57·0	1·9
April - - -	28·652	·047	·196	61·9	59·7	74·8	53·2	1·7
May - - -	28·669	·039	·236	57·4	55·6	69·4	50·0	2·5
June* - - -	28·675	·031	·154	55·4	54·3	66·7	46·8	4·7
July - - -	28·684	·047	·303	55·2	53·2	68·0	46·8	1·5
August - - -	28·728	·047	·441	54·5	52·5	67·3	45·7	1·6
September - - -	28·701	·047	·300	57·9	54·5	73·4	49·5	1·4
October - - -	28·709	·063	·299	62·2	59·5	76·1	53·8	2·3
November - - -	28·694	·055	·252	65·0	61·9	78·8	56·1	2·9
December - - -	28·659	·041	·256	68·4	66·2	82·0	59·4	2·2
Year - -	28·675	·047	—	61·8	59·6	†86·0	38·3†	2·0

* June 1867 wanting.
† These are the absolute maximum and minimum temperature of the two years.

COQUIMBO.

Latitude, 29° 56′ S. Longitude, 71° 17′ W. Height above the sea, 59 feet.

Months.	Mean at 32° F.	Barometer.		Temperature of Air.		
		Mean Daily Range.	Mean Monthly Range.	Average.	Mean Maximum.	Mean Minimum.
January	29·863	0·032	·213	65·3	75·0	58·8
February	29·863	·016	·177	66·0	75·0	59·5
March	29·882	·024	·177	62·6	70·9	56·1
April	29·949	·035	·295	60·4	68·9	54·3
May	29·945	·032	·221	56·5	66·6	47·8
June	29·957	·026	·205	53·6	66·2	45·5
July	29·957	·004	·391	53·4	61·9	46·0
August	30·020	·006	·386	55·8	64·6	47·3
September	29·978	·016	·303	58·3	66·4	50·0
October	29·973	·024	·236	59·9	67·1	52·0
November	29·934	·020	·142	61·5	70·3	54·0
December	29·875	·024	·209	63·7	72·0	56·8
Averages for 2 years.	29·934	—	—	5·97 3	—	—

The barometer observations are for the years 1849 and 1850. The temperature observations are for 1852 to 1854, ending August, with which those for September to December 1859 have been incorporated, to complete the three years. Hours, 9 a.m., 3, and 10 p.m.

LONDON:
Printed by GEORGE E. EYRE and WILLIAM SPOTTISWOODE,
Printers to the Queen's most Excellent Majesty.
For Her Majesty's Stationery Office.
[4022.—500.—7/71.]
[P. 1065.—500.—7/71.]

CHART OF METEOROLOGICAL DATA
FOR CAPE HORN & THE WEST COAST OF SOUTH AMERICA.

JANUARY.

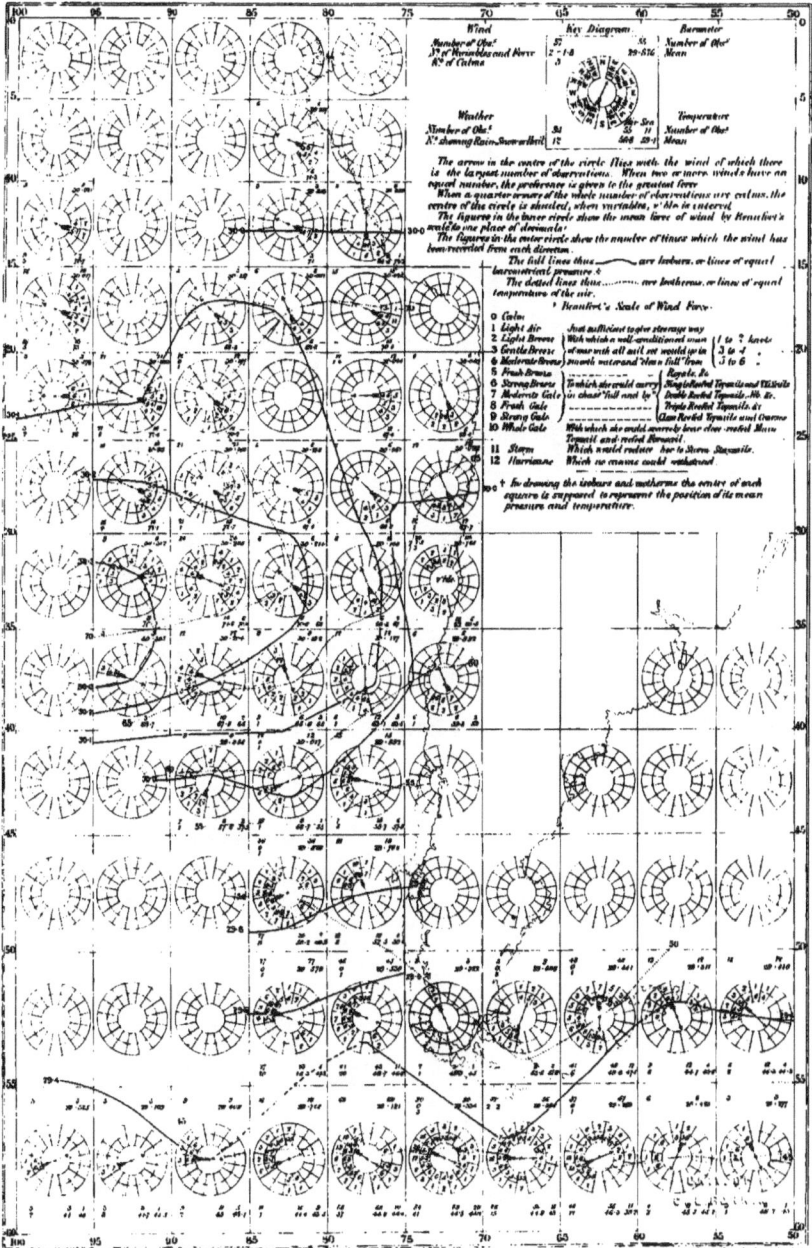

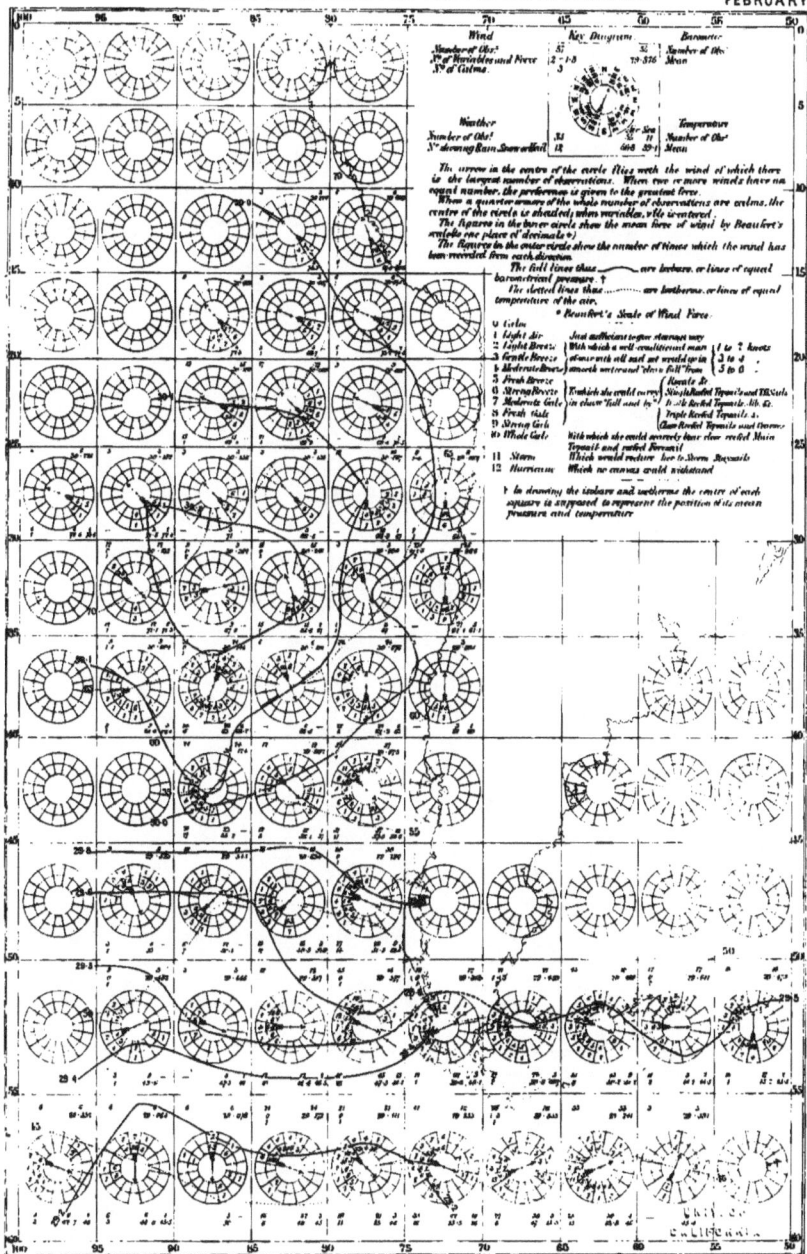

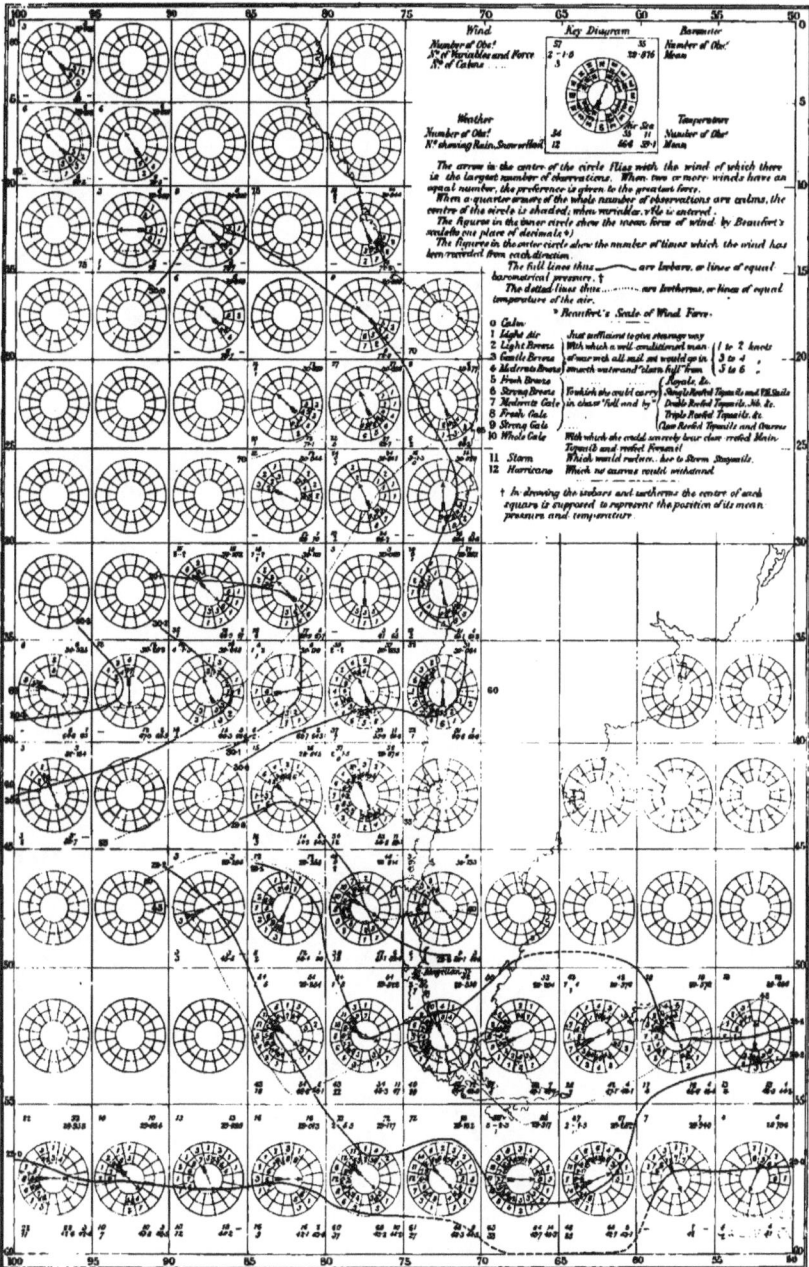

CHART OF METEOROLOGICAL DATA
FOR CAPE HORN & THE WEST COAST OF SOUTH AMERICA.

MARCH.

Wind
Number of Obs.¹
Nᵒ of Variables and Force
Nᵒ of Calms

Key Diagram

Barometer
Number of Obs.
Mean

Weather
Number of Obs.
Nᵒ showing Rain, Snow or Hail

Temperature
Number of Obs.
Mean

The arrow is the centre of the circle flies with the wind of which there is the largest number of observations. When two or more winds have an equal number, the preference is given to the greatest force.

When a quarter or more of the whole number of observations are calms, the centre of the circle is shaded; when variable, *V* is entered.

The figures in the inner circle shew the mean force of wind, by Beaufort's scale (to one place of decimals.)

The figures in the outer circle shew the number of times which the wind has been recorded from each direction.

The full lines thus ——— are Isobars, or lines of equal barometrical pressure. †

The dotted lines thus ········· are Isotherms, or lines of equal temperature of the air.

* Beaufort's Scale of Wind Force.

0 Calm
1 Light Air   Just sufficient to give steerage way
2 Light Breeze   With which a well conditioned man   { 1 to 2 knots
3 Gentle Breeze   of war with all sail set would go in   { 3 to 4 "
4 Moderate Breeze   smooth water and "clean full" from   { 5 to 6 "
5 Fresh Breeze
6 Strong Breeze   To which she could carry
7 Moderate Gale   in chase "full and by"
8 Fresh Gale
9 Strong Gale
10 Whole Gale   With which she could scarcely bear close reefed Main Topsails and reefed Foresail.
11 Storm   Which would reduce her to Storm Staysails.
12 Hurricane   Which no canvas could withstand.

† In drawing the isobars and isotherms the centre of each square is supposed to represent the position of its mean pressure and temperature.

CHART OF METEOROLOGICAL DATA
FOR CAPE HORN & THE WEST COAST OF SOUTH AMERICA.

MAY.

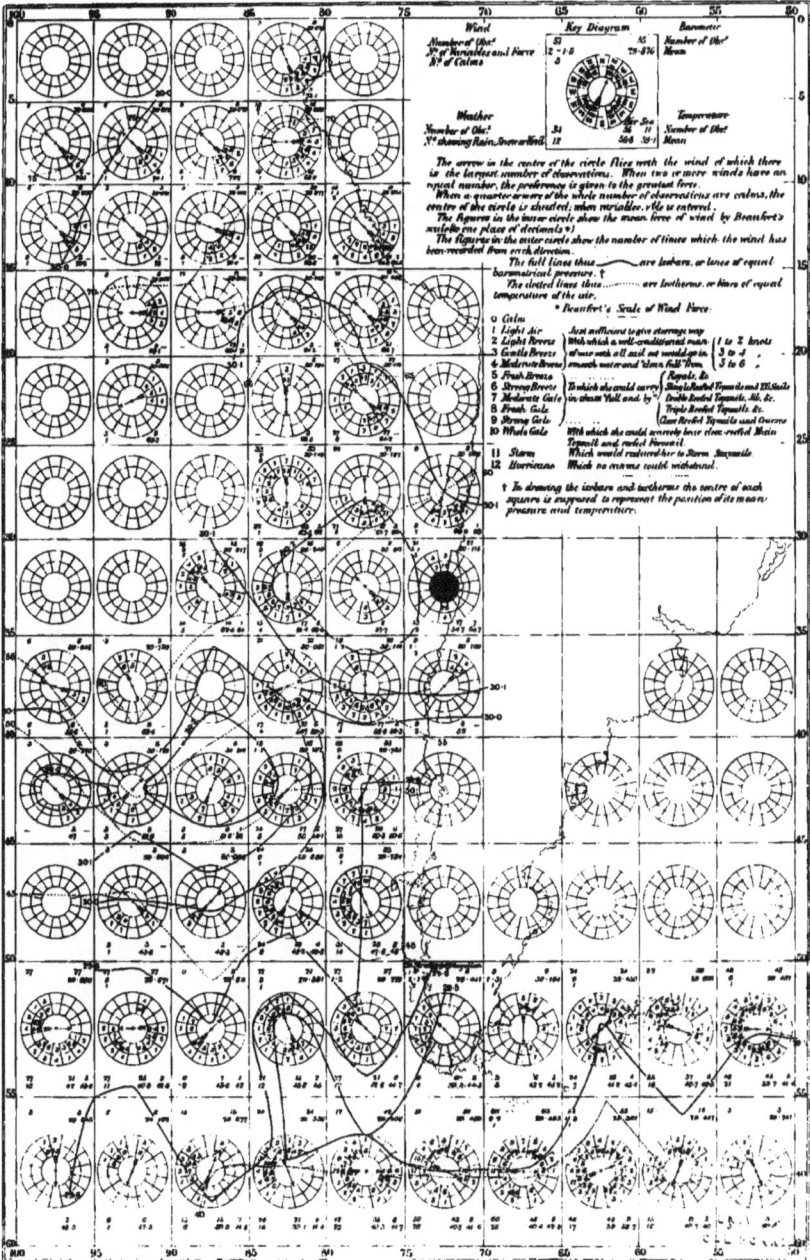

CHART OF METEOROLOGICAL DATA
FOR CAPE HORN & THE WEST COAST OF SOUTH AMERICA.

JULY.

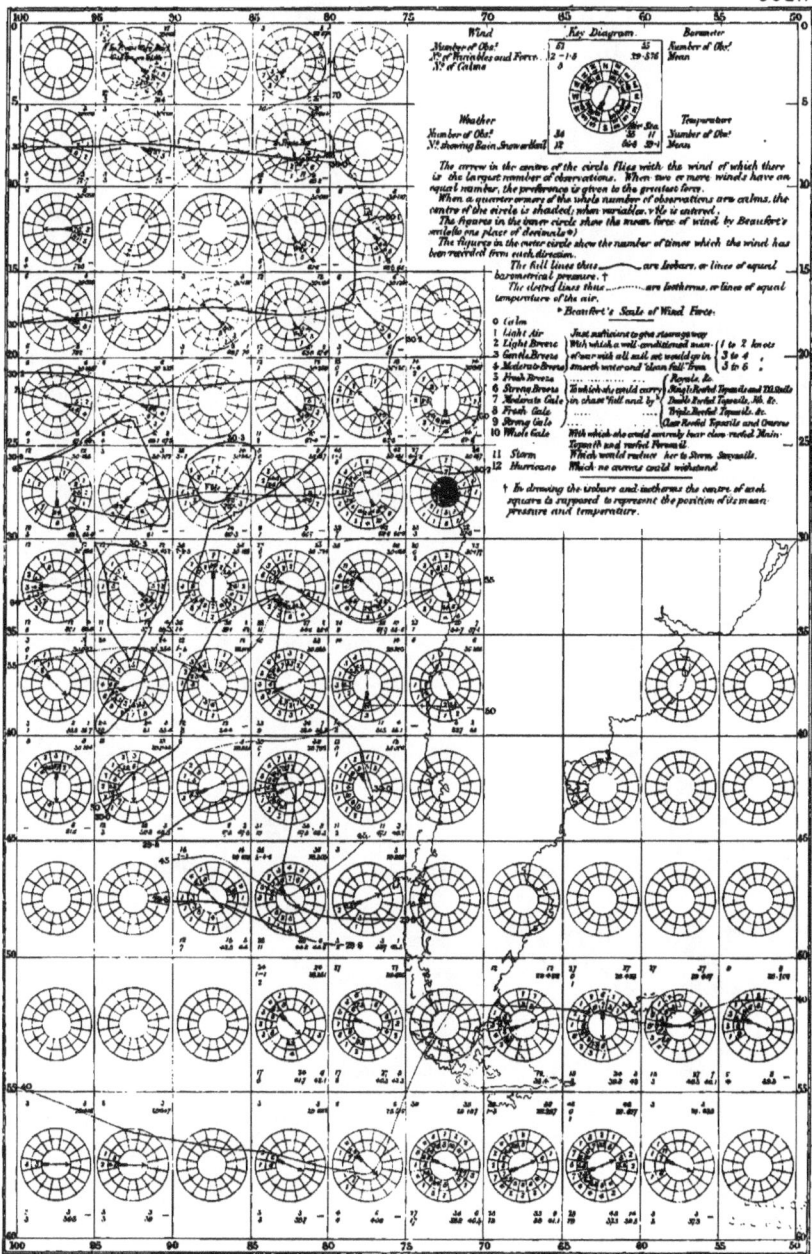

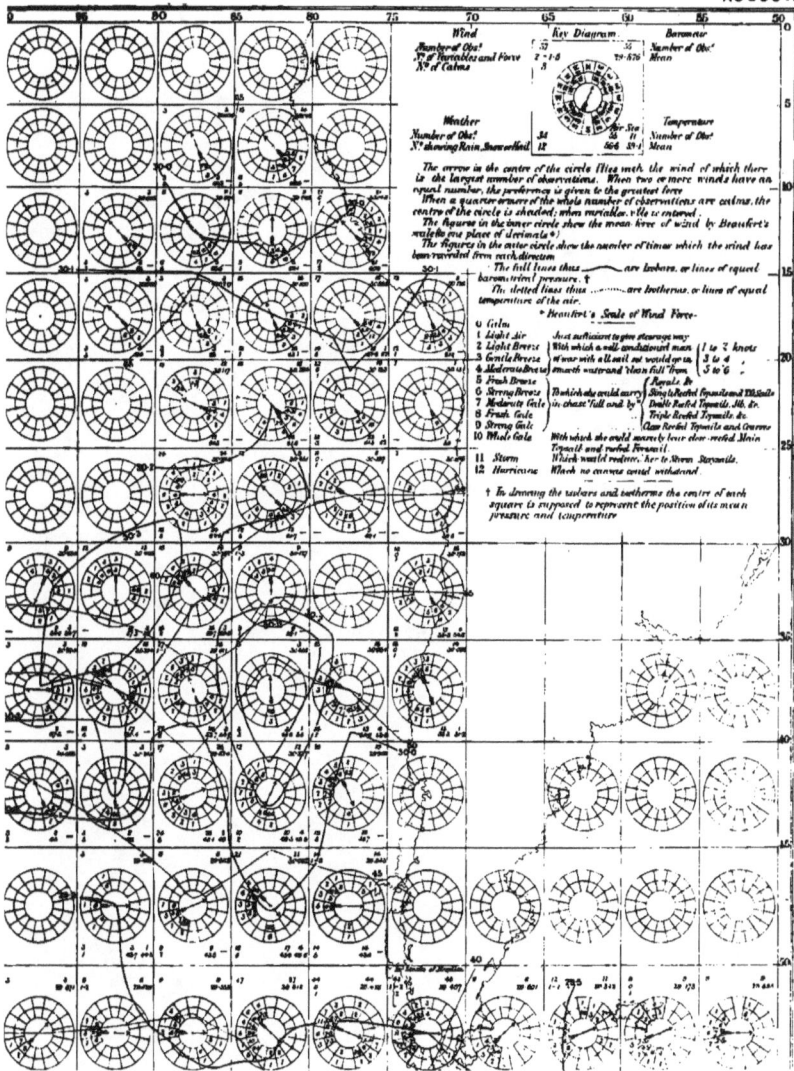

CHART OF METEOROLOGICAL DATA
FOR CAPE HORN & THE WEST COAST OF SOUTH AMERICA.

AUGUST.

Wind   Key Diagram   Barometer

The arrow in the centre of the circle flies with the wind of which there
is the largest number of observations. When two or more winds have an
equal number, the preference is given to the greatest force.
When a quarter or more of the whole number of observations are calms, the
centre of the circle is shaded; when variables, v.ble is entered.
The figures in the outer circle show the mean force of wind by Beaufort's
scale to one place of decimals (*).
The figures in the outer circle show the number of times which the wind has
been traveled from each direction.

The full lines thus ———— are Isobars, or lines of equal
barometrical pressure. †
The dotted lines thus ·········· are Isotherms, or lines of equal
temperature of the air.

* Beaufort's Scale of Wind Force.

0 Calm
1 Light Air
2 Light Breeze
3 Gentle Breeze
4 Moderate Breeze
5 Fresh Breeze
6 Strong Breeze
7 Moderate Gale
8 Fresh Gale
9 Strong Gale
10 Whole Gale
11 Storm
12 Hurricane

† In drawing the isobars and isotherms the centre of each
square is supposed to represent the position of its mean
pressure and temperature.

CHART OF METEOROLOGICAL DATA
FOR CAPE HORN & THE WEST COAST OF SOUTH AMERICA.

SEPTEMBER.

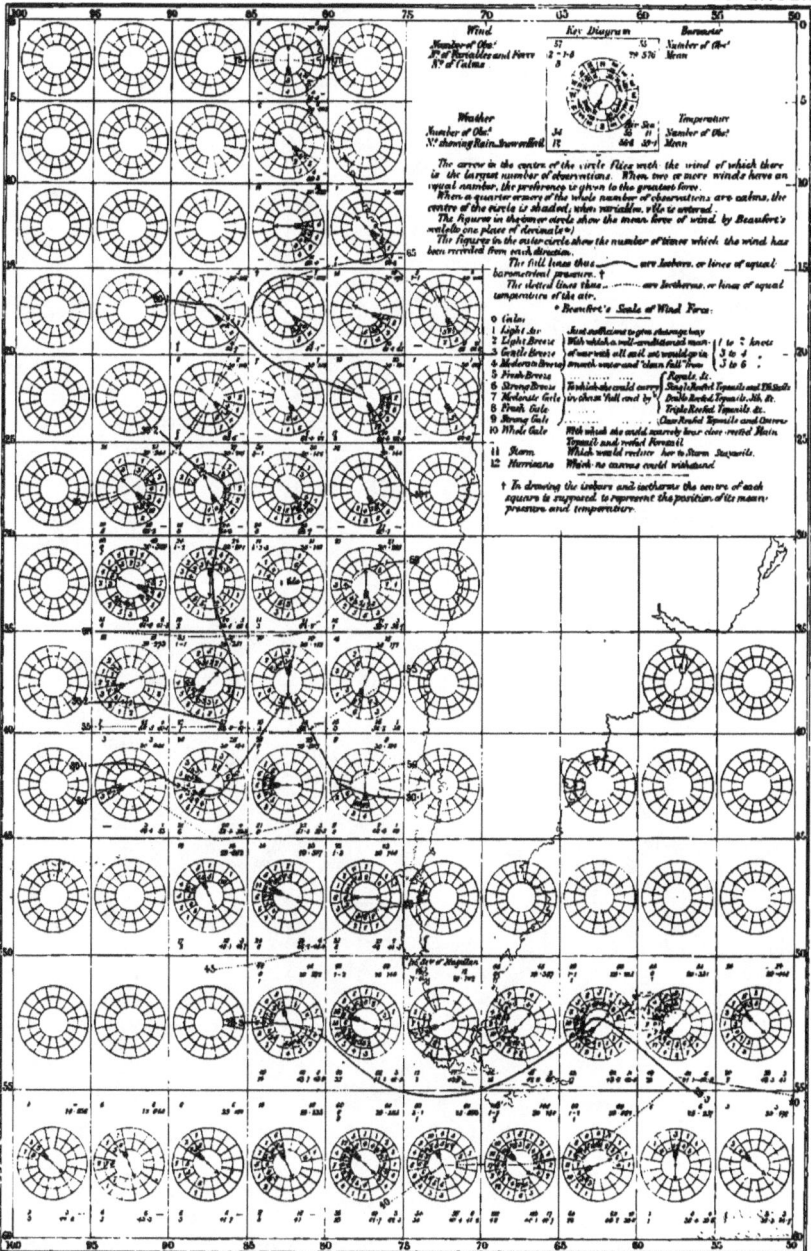

Wind

Number of Obs.
No. of Variables and Force
No. of Calms

Key Diagram

Barometer

Number of Obs.
Mean

Weather

Number of Obs.
No. showing Rain, Snow or Hail.

Temperature

Number of Obs.
Mean

The arrow in the centre of the circle flies with the wind of which there is the largest number of observations. When two or more winds have an equal number, the preference is given to the greatest force.

When a quarter or more of the whole number of observations are calms, the centre of the circle is shaded; when variable, it is watered.

The figures in the inner circle show the mean force of wind by Beaufort's notation: one place of decimals(*)

The figures in the outer circle show the number of times which the wind has been recorded from each direction.

The full lines thus _____ are isobars, or lines of equal barometrical pressure. †

The dotted lines thus _____ are isotherms, or lines of equal temperature of the air.

* Beaufort's Scale of Wind Force:

0	Calm	
1	Light Air	Just sufficient to give steerage way
2	Light Breeze	With which a well-conditioned man-
3	Gentle Breeze	of war with all sail set would go in
4	Moderate Breeze	smooth water and 'clean full' from
5	Fresh Breeze		Royals, &c.
6	Strong Breeze	To which she could carry	Single Reefed Topsails and Top Gallt. Sails
7	Moderate Gale	in chase full and by	Double Reefed Topsails, Jib, &c.
8	Fresh Gale		Triple Reefed Topsails &c.
9	Strong Gale		Close Reefed Topsails and Courses
10	Whole Gale	With which she could scarcely bear close reefed Main Topsail and reefed Foresail
11	Storm	Which would reduce her to Storm Staysails.
12	Hurricane	Which no canvas could withstand

† In drawing the isobars and isotherms the centre of each square is supposed to represent the position of its mean pressure and temperature.

Malby & Sons, Lith.

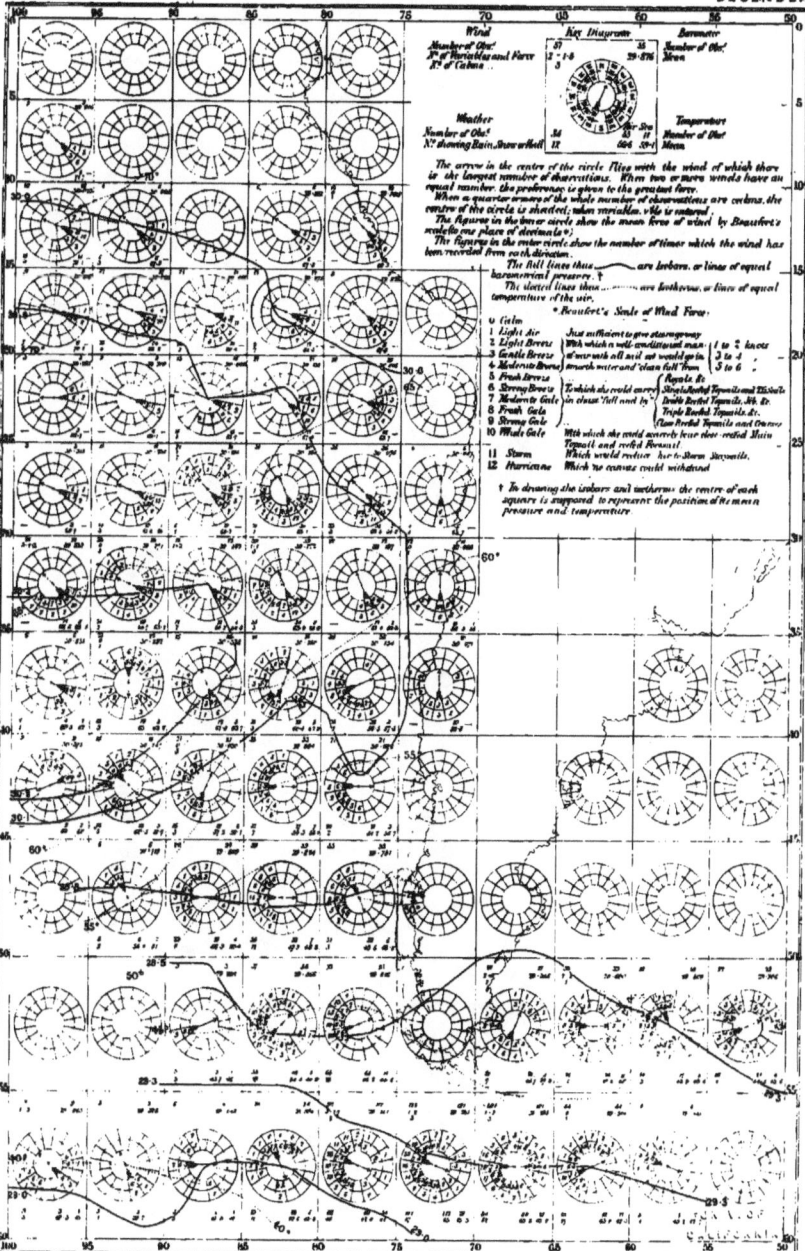